Advanced Courses in Mathematics
CRM Barcelona

Centre de Recerca Matemàtica

Managing Editor:
David Romero i Sànchez

More information about this series at http://www.springer.com/series/5038

Ana Hurtado • Steen Markvorsen • Maung Min-Oo
Vicente Palmer

Global Riemannian Geometry: Curvature and Topology

Second Edition

 Birkhäuser

Ana Hurtado
Departamento de Geometría y Topología
Universidad de Granada
Granada, Spain

Maung Min-Oo
Department of Mathematics & Statistics
McMaster University
Hamilton, ON, Canada

Steen Markvorsen
DTU Compute
Technical University of Denmark
Kgs. Lyngby, Denmark

Vicente Palmer
Departament de Matemàtiques
Universitat Jaume I
Castelló de la Plana, Spain

ISSN 2297-0304   ISSN 2297-0312   (electronic)
Advanced Courses in Mathematics - CRM Barcelona
ISBN 978-3-030-55292-3   ISBN 978-3-030-55293-0   (eBook)
https://doi.org/10.1007/978-3-030-55293-0

This book is published under the imprint Birkhäuser, www.birkhauser-science.com by the registered company Springer Nature Switzerland AG
The registered company address is: Gewerbestrasse 11, 6330 Cham, Switzerland

Contents

1. Distance Geometric Analysis on Manifolds
Ana Hurtado, Steen Markvorsen, and Vicente Palmer

2. The Dirac Operator in Geometry and Physics
Maung Min-Oo

Preface to the first edition, 2003

In July 2001, the Centre de Recerca Matematica organized at the Universitat Jaume I, in Castello de la Plana, a 20 hour Advanced Course on global Riemannian geometry: curvature and topology.

We focused our talks on two main topics:

a) The comparison theory for distance functions in spaces which have well-defined bounds on their curvature. In this setting we obtained information about diffusion processes, isoperimetric inequalities, transience, and effective resistance of the spaces in question.

b) The study of scalar curvature and positive mass theorems using spinors and the Dirac operator. After the course was finished we extended and smoothed out the material presented in the lectures, and integrated it with the background material furnished to the participants and with their many interesting comments.

It is our great pleasure to thank Professor Vicente Palmer, the local organisers Ximo Gual and Ana Lluch, the Centre de Recerca Matematica (CRM) and the Universitat Jaume I for the invitation, the cordial hospitality, the favourable planning of the course and the stimulating working atmosphere.

Preface to the second edition, 2020

In this second edition we update and extend the 2003-edition of the report in a number of ways.

Our good friends and long-time co-authors Ana Hurtado and Vicente Palmer have contributed significantly to the first chapter.

There we have now embedded short surveys on a number of recent extensions – including, for example, promising results concerning the geometry of exit time moment spectra and concerning potential analysis in weighted Riemannian manifolds, to mention but two of the new added features.

In the second chapter Maung Min-Oo has added an update and an epilogue discussing recent developments related to the topics of his lectures – including, for example, results pertaining to an early conjecture of his on the geometry of the scalar curvature and speculations on new geometric approaches to the Index Theorem.

It is our great pleasure to thank Dorothy Mazlum and Birkhäuser for suggesting this second edition of the CRM notes.

Distance Geometric Analysis on Manifolds

Ana Hurtado, Steen Markvorsen, and Vicente Palmer

1. Appetizer and Introduction

It is a natural and indeed a classical question to ask: "What is the effective resistance of, say, a hyperboloid or a helicoid if the surface is made of a homogeneous conducting material?".

In these notes we will study the precise meaning of this and several other related questions and analyze how the answers depend on the curvature and topology of the given surfaces and manifolds. We will focus mainly on minimal submanifolds in ambient spaces which are assumed to have a well-defined upper (or lower) bound on their sectional curvatures.

One key ingredient is the comparison theory for distance functions in such spaces. In particular we establish and use a comparison result for the Laplacian of geometrically restricted distance functions. It is in this setting that we obtain information about such diverse phenomena as diffusion processes, isoperimetric inequalities, Dirichlet eigenvalues, transience, recurrence, and effective resistance of the spaces in question. In this second edition of the present notes we extend those previous findings in four ways: Firstly, we include comparison results for the exit time moment spectrum for compact domains in Riemannian manifolds; Secondly, and most substantially, we report on very recent results obtained by the first and third author together with C. Rosales concerning comparison results for the capacities and the type problem (transient versus recurrent) in weighted Riemannian manifolds; Thirdly we survey how some of the purely Riemannian results on transience and recurrence can be lifted to the setting of spacelike submanifolds in Lorentzian manifolds; Fourthly, the comparison spaces that we employ for some of the new results are typically so-called model spaces, i.e., warped products (generalized surfaces of revolution) where 'all the geometry' in each case is determined by a given radial warping function and a given weight function.

1991 *Mathematics Subject Classification.* Primary 53C, 05C, 58-99, 58J, 60J.

Key words and phrases. Minimal submanifolds, Laplace equation, Poisson equation, transience, recurrence, hyperbolic type, parabolic type, capacity, weighted potential theory, isoperimetric inequalities, comparison theory, volume growth, model spaces, exit time moment spectra, heat content, Poisson hierarchy, weighted manifolds, weighted submanifolds, weighted curvature, Scherk's surface, Scherk's graph.

FIGURE 1. How large is the current through the Helicoid when the battery supplies a unit Voltage? (The boundary is assumed to be superconducting and the surface itself is assumed to be made of a homogeneous conducting material.)

In a sense, all the different phenomena that we consider are 'driven' by the Laplace operator which in turn depends on the background curvatures and the weight function. One key message of this report is that the Laplacian is a particularly 'swift' operator – for example on minimal submanifolds in ambient spaces with small sectional curvatures – but depending on the weight functions. Specifically, we observe and report new findings about this behaviour in the contexts of both Riemannian, Lorentzian, and weighted geometries, see Sections 12 and 20–27. Similar results generally hold true within the intrinsic geometry of the manifolds themselves – often even with Ricci curvature lower bounds (see, e.g., the survey [Zhu]) as a substitute for the specific assumption of a *lower bound* on sectional curvatures.

2. The Comparison Setting and Preliminaries

We consider a complete immersed submanifold P^m in a Riemannian manifold N^n, and denote by D^P and D^N the Riemannian connections of P and N, respectively. We refer to the excellent general monographs on Riemannian geometry – e.g., [Sa], [CheeE], and [Cha2] – for the basic notions, that will be applied in these notes. In particular we shall be concerned with the second-order behavior of certain functions on P which are obtained by restriction from the ambient space N as displayed in Proposition 3.1 below. The second-order derivatives are defined in terms of the Hessian operators Hess^N, Hess^P and their traces Δ^N and Δ^P, respectively (see, e.g., [Sa] p. 31). The difference between these operators quite naturally involves geometric second-order information about how P^m actually sits inside N^n. This information is provided by the *second fundamental form* α (resp. the mean curvature H) of P in N (see [Sa] p. 47). If the functions under consideration are essentially *distance* functions in N – or suitably modified distance functions – then their second-order behavior is strongly influenced by the curvatures of N, as is directly expressed by the second variation formula for geodesics ([Sa] p. 90). As is well known, the ensuing and by now classical comparison theorems for Jacobi fields give rise to the celebrated Toponogov theorems for geodesic triangles and to powerful results concerning the global structure of Riemannian spaces ([Sa], Chapters IV–V). In these notes, however, we shall mainly apply the Jacobi field comparison theory only off the *cut loci* of the ambient space N, or more precisely, within the *regular balls* of N as defined in Definition 3.4 below. On the other hand, from the point of view of a given (minimal) submanifold P in N, our results for P are semi-global in the sense that they apply to domains which are not necessarily distance-regular within P.

3. Analysis of Riemannian Distance Functions

Let $\mu : N \mapsto \mathbb{R}$ denote a smooth function on N. Then the restriction $\tilde{\mu} = \mu_{|P}$ is a smooth function on P and the respective Hessians $\mathrm{Hess}^N(\mu)$ and $\mathrm{Hess}^P(\tilde{\mu})$ are related as follows:

Proposition 3.1 ([JK] p. 713).

$$\mathrm{Hess}^P(\tilde{\mu})(X,Y) = \mathrm{Hess}^N(\mu)(X,Y) \\ + \langle \nabla^N(\mu), \alpha(X,Y) \rangle \tag{3.1}$$

for all tangent vectors $X, Y \in TP \subseteq TN$, *where* α *is the second fundamental form of* P *in* N.

Proof.

$$\begin{aligned}
\mathrm{Hess}^P(\tilde{\mu})(X,Y) &= \langle \mathrm{D}^P_X \nabla^P \tilde{\mu}, Y \rangle \\
&= \langle \mathrm{D}^N_X \nabla^P \tilde{\mu} - \alpha(X, \nabla^P \tilde{\mu}), Y \rangle \\
&= \langle \mathrm{D}^N_X \nabla^P \tilde{\mu}, Y \rangle \\
&= X\left(\langle \nabla^P \tilde{\mu}, Y \rangle \right) - \langle \nabla^P \tilde{\mu}, \mathrm{D}^N_X Y \rangle \\
&= \langle \mathrm{D}^N_X \nabla^N \mu, Y \rangle + \langle (\nabla^N \mu)^\perp, \mathrm{D}^N_X Y \rangle \\
&= \mathrm{Hess}^N(\mu)(X, Y) + \langle (\nabla^N \mu)^\perp, \alpha(X, Y) \rangle \\
&= \mathrm{Hess}^N(\mu)(X, Y) + \langle \nabla^N \mu, \alpha(X, Y) \rangle .
\end{aligned} \tag{3.2}$$

$\square$

If we modify μ to $F \circ \mu$ by a smooth function $F : \mathbb{R} \mapsto \mathbb{R}$, then we get

Lemma 3.2.

$$\mathrm{Hess}^N(F \circ \mu)(X,X) = F''(\mu) \cdot \langle \nabla^N(\mu), X \rangle^2 \\ + F'(\mu) \cdot \mathrm{Hess}^N(\mu)(X,X) , \tag{3.3}$$

for all $X \in TN^n$.

In the following we write $\mu = \tilde{\mu}$. Combining (3.1) and (3.3) then gives

Corollary 3.3.

$$
\begin{aligned}
\mathrm{Hess}^P (F \circ \mu)(X, X) = {} & F''(\mu) \cdot \langle \nabla^N(\mu), X \rangle^2 \\
& + F'(\mu) \cdot \mathrm{Hess}^N(\mu)(X, X) \\
& + \langle \nabla^N(\mu), \alpha(X, X) \rangle .
\end{aligned}
\tag{3.4}
$$

for all $X \in TP^m$.

In what follows the function μ will always be a distance function in N – either from a point p in which case we set $\mu(x) = \mathrm{dist}_N(p, x) = r(x)$, or from a totally geodesic hypersurface V^{n-1} in N in which case we let $\mu(x) = \mathrm{dist}_N(V, x) = \eta(x)$. The function F will always be chosen, so that $F \circ \mu$ is smooth inside the respective *regular* balls around p and inside the *regular* tubes around V, which we now define. The sectional curvatures of the two-planes Ω in the tangent bundle of the ambient space N are denoted by $K_N(\Omega)$, see, e.g., [Sa], Section II.3. Concerning the notation: In the following both Hess^N and Hess will be used invariantly for both the Hessian in the ambient manifold N, as well as in a purely intrinsic context where only N and not any of its submanifolds is under consideration.

Definition 3.4. The set $B_R(p) = \{x \in N \mid r(x) \le R\}$ is a *regular ball* of radius R around p if $R < \frac{\pi}{\sqrt{k}}$ and $R < i_N(p)$, where

1. $k = \sup_{x \in B_R} \{K_N(\Omega) \mid \Omega$ is a two-plane in $TB_R(x)\}$,
2. $\frac{\pi}{\sqrt{k}} = \infty$ if $k \le 0$,   and
3. $i_N(p) = $ the injectivity radius of the exponential map from p in N.

Definition 3.5. The set

$$
B_R(V) = \{x \in N \mid \eta(x) \le R\}
$$

is a *regular tube* of radius R around V if $R < \frac{\pi}{2\sqrt{k}}$ and $R < i_N(V)$.

Within regular balls we may now apply Jacobi field and index type comparison theory to display the significant sensitivity of $\mathrm{Hess}^P(F \circ \mu)(X, X)$ on the curvature of the ambient space N and on the geometry of P in N, respectively.

The mean curvature functions of the respective distance spheres in the constant curvature spaces $\mathbb{K}^m(b)$ play a significant role for these comparison results:

Lemma 3.6. *The mean curvature* $h_b(r)$ *of the geodesic sphere* $\partial B_r^{b,m}(\tilde{p}) = S_r^{b,m-1}(\tilde{p})$ *of radius* r *(and center* $\tilde{p}$ *anywhere) in the space form* $\mathbb{K}^m(b)$ *of constant curvature* b *is given by*

$$
h_b(r) = \begin{cases}
\sqrt{b} \cot(\sqrt{b}\, r), & \textit{if } b > 0 \\
1/r & \textit{if } b = 0 \\
\sqrt{-b} \coth(\sqrt{-b}\, r) & \textit{if } b < 0 .
\end{cases}
\tag{3.5}
$$

The mean curvature $f_b(\eta)$ of the geodesic distance set $\partial B_\eta^{b,m}(\tilde{V})$ of radius η (and totally geodesic hypersurface $\tilde{V}$ anywhere) in $\mathbb{K}^m(b)$ is given by

$$f_b(\eta) = \begin{cases} -\sqrt{b}\,\tan(\sqrt{b}\,\eta) & \text{if } b > 0 \\ 0 & \text{if } b = 0 \\ \sqrt{-b}\,\tanh(\sqrt{-b}\,\eta) & \text{if } b < 0\,. \end{cases} \tag{3.6}$$

Exercise 3.7. Prove this lemma.

The comparison statements for the respective Hessians (when we assume an explicit upper or lower bound on the sectional curvatures for N) are then respectively

Proposition 3.8. *If $K_N \leq b$ (respectively $K_N \geq b$), $b \in \mathbb{R}$, then we have for every unit vector X in the tangent bundle of the regular ball $B_R(p) - p$ in N:*

$$\text{Hess}^N(r)(X, X) \geq (\text{resp. } \leq)\ h_b(r) \cdot \left(1 - \langle \nabla^N r, X \rangle^2\right)\,. \tag{3.7}$$

And correspondingly for the Hessian of the distance to a given totally geodesic hypersurface V:

Proposition 3.9. *If $K_N \leq b$ (respectively $K_N \geq b$), $b \in \mathbb{R}$, then we have for every unit vector in the tangent bundle of the regular tube $B_R(V) - V$ in N:*

$$\text{Hess}^N(\eta)(X, X) \geq (\text{resp. } \leq)\ f_b(\eta) \cdot \left(1 - \langle \nabla^N \eta, X \rangle^2\right)\,. \tag{3.8}$$

Proof. We consider the latter (tube) case. In $B_R(V)$ the distance function η is well defined and smooth outside V itself by assumption of regularity. We let $T = \nabla^N(\eta)$ and consider a point $q \in B_R(V) - V$. Then $T(q) = \dot{\gamma}(\eta(q))$, where $\gamma : [0, \eta(q)] \mapsto N$ is the unique unit speed minimal geodesic from V to q. Now let X be a unit vector in the tangent space $T_q(N)$ and let $X(s)$ denote the Jacobi vector field along γ generated in the usual way by X through minimal geodesics connecting to V such that $X(\eta(q)) = X$ and such that $X'(0) = \text{D}_{\dot{\gamma}(0)}^N = 0$. Then it follows from, e.g., [CheeE] pp. 20–21, that

$$\text{Hess}^N(\eta)_{|_q}(X, X) = I_\gamma(X, X) - \int_0^\eta (T\langle X, T\rangle)^2\, ds\,, \tag{3.9}$$

where I_γ is the index form along γ. If we define

$$X^\perp(s) = X(s) - \langle T(s), X(s)\rangle \cdot T(s)\,, \tag{3.10}$$

so that

$$\left|\nabla_T^N X\right|^2 = \left|\nabla_T^N X^\perp\right|^2 + (T\langle X, T\rangle)^2\,, \tag{3.11}$$

we finally get

$$\text{Hess}^N(\eta)_{|_q}(X, X) = I_\gamma(X^\perp, X^\perp) \tag{3.12}$$

for all $X \in T_q N$. Standard index comparison theory now gives Proposition 3.9, as in, e.g., [CheeE] or [Sa]. Proposition 3.8 follows verbatim. $\qquad\square$

Combining these results now gives estimates on $\text{Hess}^P(F \circ r)(X, X)$ and $\text{Hess}^P(F \circ \eta)(X, X)$ for all unit X in the tangent bundle of the submanifold $P \cap B_R(p)$:

Proposition 3.10. *Let $P^m \subseteq N^n$ denote a submanifold (or the image of an isometric immersion) into a regular ball $B_R(p)$ or into a regular tube $B_R(V)$ around a totally geodesic hypersurface V in N. Assume that the function $F \circ r$, respectively $F \circ \eta$, is a smooth function on P within the regular domain. Suppose that*

$$\begin{cases} K_N & \leq b \text{ for some } b \in \mathbb{R}, \text{ and that} \\ \frac{d}{dt}F(t) & \geq 0 \text{ for all } t \in [0, R]. \end{cases} \tag{3.13}$$

Then we have for every tangent unit vector X to P in the regular domain

$$\begin{aligned}
\text{Hess}^P(F \circ r)(X, X) \geq \ & (F''(r) - F'(r)h_b(r)) \cdot \langle \nabla^N r, X \rangle^2 \\
& + F'(r) \cdot (h_b(r) + \langle \nabla^N r, \alpha(X, X) \rangle),
\end{aligned} \tag{3.14}$$

and

$$\begin{aligned}
\text{Hess}^P(F \circ \eta)(X, X) \geq \ & (F''(\eta) - F'(\eta)f_b(\eta)) \cdot \langle \nabla^N r, X \rangle^2 \\
& + F'(\eta) (f_b(\eta) + \langle \nabla^N r, \alpha(X, X) \rangle).
\end{aligned} \tag{3.15}$$

There are two important corollaries to be noted:

Corollary 3.11. *If precisely one of the inequalities in the assumptions (3.13) is reversed, then the inequalities (3.14) and (3.15) are likewise reversed.*

Corollary 3.12. *If at least one of the inequalities in the assumptions (3.13) is actually an equality (i.e., $N = \mathbb{K}^n(b)$, or $F(t) = $ constant), then the inequalities (3.14) and (3.15) are equalities as well.*

Furthermore, if we trace both sides of the Hessian inequalities over an orthonormal basis $\{ X_1, \ldots, X_m \}$ of the tangent space $T_q P$ of P at some given point $q \in P$, then we get the corresponding inequalities for the Laplacian on P acting on the functions $F \circ r$ and $F \circ \eta$ respectively:

Proposition 3.13. *Suppose again that the assumptions (3.13) are satisfied. Then, respectively*

$$\begin{aligned}
\Delta^P(F \circ r) \geq \ & (F''(r) - F'(r)h_b(r)) \cdot |\nabla^P r|^2 \\
& + mF'(r) \cdot (h_b(r) + \langle \nabla^N r, H \rangle),
\end{aligned} \tag{3.16}$$

and

$$\begin{aligned}
\Delta^P(F \circ \eta) \geq \ & (F''(\eta) - F'(\eta)f_b(\eta)) \cdot |\nabla^P \eta|^2 \\
& + mF'(\eta) \cdot (f_b(\eta) + \langle \nabla^N \eta, H \rangle),
\end{aligned} \tag{3.17}$$

where H denotes the mean curvature vector of P in N.

Again, because of their importance for the further developments in these notes, we state the following consequences obtained by reversing the inequalities in the assumptions on the sectional curvature bound and on the sign of F', respectively.

Corollary 3.14. *If precisely one of the inequalities in the assumptions (3.13) is reversed, then the inequalities (3.16) and (3.17) are likewise reversed.*

Corollary 3.15. *If at least one of the inequalities in the assumptions (3.13) is actually an equality (i.e., $N = \mathbb{K}^n(b)$ or $F(t) = $ constant), then the inequalities (3.16) and (3.17) are clearly equalities as well.*

In particular we have the following

Corollary 3.16. *Let $P^m \subseteq \mathbb{K}^n(b)$ denote a submanifold (or the image of an isometric immersion of P^m) into a regular ball $B_R^{b,n}(\tilde{p})$ or into a regular tube $B_R^{b,n}(\tilde{V})$ in the space form $\mathbb{K}^n(b)$ of constant curvature b. Then*

$$\Delta^P(F \circ r) = (F''(r) - F'(r)h_b(r)) \cdot |\nabla^P r|^2$$
$$+ mF'(r) \cdot (h_b(r) + \langle \nabla^N r, H \rangle), \tag{3.18}$$

and

$$\Delta^P(F \circ \eta) = (F''(\eta) - F'(\eta)f_b(\eta)) \cdot |\nabla^P \eta|^2$$
$$+ mF'(\eta) \cdot (f_b(\eta) + \langle \nabla^N \eta, H \rangle) \tag{3.19}$$

– independently of the possible sign changes of $F'(t)$ for $t \in [0, R]$.

The information encoded in these results is particularly transparent for minimal submanifolds N, i.e., for $H = 0$ at all points in $D_R(p) = N \cap B_R(p)$, and even more so under the extra condition that

$$F''(r) - F'(r)h_b(r) = 0, \tag{3.20}$$

or

$$F''(\eta) - F'(\eta)f_b(\eta) = 0, \tag{3.21}$$

respectively. We now consider this situation in some detail.

The complete solution to (3.20) is

$$F(r) = \begin{cases} c_1 + c_2 \cos(\sqrt{b}\,r) & \text{for } b > 0 \\ c_1 + c_2\, r^2 & \text{for } b = 0 \\ c_1 + c_2 \cosh(\sqrt{-b}\,r) & \text{for } b < 0, \end{cases} \tag{3.22}$$

and the complete solution to (3.21) is

$$F(\eta) = \begin{cases} c_1 + c_2 \sin(\sqrt{b}\,\eta) & \text{for } b > 0 \\ c_1 + c_2\, \eta & \text{for } b = 0 \\ c_1 + c_2 \sinh(\sqrt{-b}\,\eta) & \text{for } b < 0, \end{cases} \tag{3.23}$$

where c_1 and c_2 are arbitrary constants of integration.

Among these solutions we find in particular the following ones, which may be expressed in terms of the generalized *sine*-function $Q_b(t)$:

$$F_b^p(r) = \int Q_b(r)\, dr = \begin{cases} -\frac{1}{b} \cos(\sqrt{b}\,r) & \text{for } b > 0 \\ \frac{1}{2}r^2 & \text{for } b = 0 \\ -\frac{1}{b} \cosh(\sqrt{-b}\,r) & \text{for } b < 0, \end{cases} \tag{3.24}$$

and

$$
F_b^V(\eta) = Q_b(\eta) = \begin{cases} \frac{1}{\sqrt{b}} \sin(\sqrt{b}\,\eta) & \text{for } b > 0 \\ \eta & \text{for } b = 0 \\ \frac{1}{\sqrt{-b}} \sinh(\sqrt{-b}\,\eta) & \text{for } b < 0 \,. \end{cases} \tag{3.25}
$$

These functions are eigenfunctions of the Laplacian for every minimal submanifold P^m in the respective space forms $\mathbb{K}^n(b)$ whenever $b \neq 0$. The corresponding 'free' eigenvalues (of $-\Delta^P$) are $m\,b$:

Proposition 3.17 ([Ma5]). *Let P^m be a minimal submanifold (or just a minimal isometric immersion of P^m) in $\mathbb{K}^n(b)$. Then*

$$
\Delta^P(F_b^p(r)) = \begin{cases} -m\,b\,F_b^p(r)\,, & \text{for } b \neq 0 \\ m & \text{for } b = 0\,, \end{cases} \tag{3.26}
$$

and

$$
\Delta^P(F_b^V(\eta)) = \begin{cases} -m\,b\,F_b^V(\eta)\,, & \text{for } b \neq 0 \\ 0 & \text{for } b = 0\,. \end{cases} \tag{3.27}
$$

Proof. This follows directly by inserting the F-functions into the equations (3.18) and (3.19). □

We observe that the minimal submanifolds of the Euclidean spaces are quite special in this context. Instead of supporting eigenfunctions (of the type considered here), they support harmonic functions and functions satisfying the Poisson problem $\Delta^P \psi = m$.

The former functions are, of course, very important for the study of potential theory on the manifolds in question; the latter functions are likewise related to the study of the mean time of first exit for the Brownian motion from domains of the manifolds.

For these reasons – among others – we will study such problems and their solutions, i.e., eigenfunctions, harmonic functions and exit time functions in the much more general setting of minimal submanifolds P^m in ambient spaces N^n of variable curvature.

And this is precisely what the setting of this section and in particular the Laplace comparison Proposition 3.13 together with the observations in Corollaries 3.14 and 3.15 turn out to be good for.

4. Analysis of Lorentzian Distance Functions

For comparison, and before going further into the Riemannian setting, we briefly present the corresponding Hessian analysis of the distance function from a point in a Lorentzian manifold and its restriction to a spacelike hypersurface. The results can be found in [AHP], where the corresponding Hessian analysis was also carried out, i.e., the analysis of the Lorentzian distance from an achronal spacelike hypersurface in the style of Proposition 3.9. Recall that in Section 3 we also considered the analysis of the distance from a totally geodesic hypersurface P in the ambient Riemannian manifold N.

Let (N^{n+1}, g) denote an $(n+1)$-dimensional spacetime, that is, a time-oriented Lorentzian manifold of dimension $n + 1 \geq 2$. The metric tensor g has index 1 in this case, and, as we did in the Riemannian context, we shall denote it alternatively as $g = \langle\,,\,\rangle$ (see, e.g., [O'N] as a standard reference for this section).

Given p, q two points in N, one says that q is in the chronological future of p, written $p \ll q$, if there exists a future-directed timelike curve from p to q. Similarly, q is in the causal future of p, written $p < q$, if there exists a future-directed causal (i.e., nonspacelike) curve from p to q.

Then the chronological future $I^+(p)$ of a point $p \in N$ is defined as

$$I^+(p) = \{q \in N : p \ll q\} \,.$$

The Lorentzian distance function $d : N \times N \to [0, +\infty]$ for an arbitrary spacetime may fail to be continuous in general, and may also fail to be finite-valued. But there are geometric restrictions that guarantee a good behavior of d. For example, globally hyperbolic spacetimes turn out to be the natural class of spacetimes for which the Lorentzian distance function is finite-valued and continuous.

Given a point $p \in N$, one can define the Lorentzian distance function $d_p : M \to [0, +\infty]$ with respect to p by

$$d_p(q) = d(p, q) \,.$$

In order to guarantee the smoothness of d_p, we need to restrict this function on certain special subsets of N. Let $T_{-1}N|_p$ be the following set

$$T_{-1}N|_p = \{v \in T_pN : v \text{ is a future-directed timelike unit vector}\} \,.$$

Define the function $s_p : T_{-1}N|_p \to [0, +\infty]$ by

$$s_p(v) = \sup\{t \geq 0 : d_p(\gamma_v(t)) = t\} \,,$$

where $\gamma_v : [0, a) \to N$ is the future inextendible geodesic starting at p with initial velocity v. Then we define

$$\tilde{I}^+(p) = \{tv : \text{ for all } v \in T_{-1}N|_p \text{ and } 0 < t < s_p(v)\}$$

and consider the subset $\mathcal{I}^+(p) \subset N$ given by

$$\mathcal{I}^+(p) = \exp_p(\mathrm{int}(\tilde{I}^+(p))) \subset I^+(p) \,.$$

Observe that the exponential map

$$\exp_p : \mathrm{int}(\tilde{I}^+(p)) \to \mathcal{I}^+(p)$$

is a diffeomorphism and $\mathcal{I}^+(p)$ is an open subset (possible empty).

Remark 4.1. When $b \geq 0$, the Lorentzian space form of constant sectional curvature b, which we denote as N_b^{n+1}, is globally hyperbolic and geodesically complete, and every future directed timelike unit geodesic γ_b in N_b^{n+1} realizes the Lorentzian distance between its points. In particular, if $b \geq 0$ then $\mathcal{I}^+(p) = I^+(p)$ for every point $p \in N_b^{n+1}$ (see [EGK, Remark 3.2]).

Let $\psi : \Sigma^n \to N^{n+1}$ be a spacelike hypersurface immersed into the spacetime N. Since N is time-oriented, there exists a unique future-directed timelike unit normal field ν globally defined on Σ. We will refer to ν as the future-directed Gauss map of Σ. Let A denote the shape operator of Σ with respect to ν. The function $H = -(1/n)\mathrm{tr}(A)$ defines the future mean curvature of Σ. The choice of the sign $-$ in our definition of H is motivated by the fact that in that case the mean curvature vector is given by $\overrightarrow{H} = H\nu$. Therefore, $H(p) > 0$ at a point $p \in \Sigma$ if and only if $\overrightarrow{H}(p)$ is future-directed.

Then we have the following result, which is a Lorentzian version of Proposition 3.1, when we consider the Lorentzian distance and its restriction to the spacelike hypersurface Σ:

Proposition 4.2 ([AHP]). *Let us assume that there exists a point $p \in N$ such that $\mathcal{I}^+(p) \neq \emptyset$ and that $\psi(\Sigma) \subset \mathcal{I}^+(p)$. Let $r = d_p$ denote the Lorentzian distance function from p, and let $u = r \circ \psi : \Sigma \to (0, \infty)$ be the function r along the hypersurface, which is a smooth function on Σ. Then, we have that*

$$\mathrm{Hess}^{\Sigma}(u)(X, X) = \mathrm{Hess}^N(r)(X, X) - \langle AX, X \rangle \sqrt{1 + |\nabla^{\Sigma} u|^2} \qquad (4.1)$$

for all tangent vector $X \in T\Sigma \subseteq TN$.

Taking traces we get:

$$\Delta^{\Sigma} u = \Delta^N r + \mathrm{Hess}^N(r)(\nu, \nu) + nH\sqrt{1 + |\nabla^{\Sigma} u|^2} , \qquad (4.2)$$

where $\Delta^N r$ is the (Lorentzian) Laplacian of r and $H = -(1/n)\mathrm{tr}(A)$ is the mean curvature of Σ.

Proof. Along Σ we have that

$$\nabla^N r = \nabla^{\Sigma} u - \langle \nabla^N r, \nu \rangle \nu ,$$

where $\nabla^{\Sigma} u$ denotes the gradient of u on Σ. Using that $\langle \nabla^N r, \nabla^N r \rangle = -1$ and $\langle \nabla^N r, \nu \rangle > 0$, we have that

$$\langle \nabla^N r, \nu \rangle = \sqrt{1 + |\nabla^{\Sigma} u|^2} \geq 1 , \quad \text{so that} \quad \nabla^N r = \nabla^{\Sigma} u - \sqrt{1 + |\nabla^{\Sigma} u|^2}\nu .$$

Moreover, from the Gauss and Weingarten formulae, we get

$$\nabla^N_X \nabla^N r = \nabla^{\Sigma}_X \nabla^{\Sigma} u + \sqrt{1 + |\nabla^{\Sigma} u|^2}AX + \langle AX, \nabla^{\Sigma} u \rangle \nu - X(\sqrt{1 + |\nabla^{\Sigma} u|^2})\nu$$

for every tangent vector field $X \in T\Sigma$. Thus,

$$\mathrm{Hess}^{\Sigma}(u)(X, X) = \mathrm{Hess}^N(r)(X, X) - \langle AX, X \rangle \sqrt{1 + |\nabla^{\Sigma} u|^2} \qquad (4.3)$$

for all tangent vector $X \in T\Sigma \subseteq TN$. $\qquad \square$

Let us consider now the Lorentzian R-spheres in the Lorentzian space form N_b^{n+1} (when $\mathcal{I}^+(p) \neq \emptyset$), which are defined as the level sets

$$\Sigma_b(R) = \{q \in \mathcal{I}^+(p) : d_p(q) = R\} \subset N_b^{n+1} .$$

As in the Riemannian case, we have the following

Lemma 4.3 ([AHP]). *The mean curvature $f_b(r)$ of the geodesic sphere $\Sigma_b(r)$ of radius r (and center p anywhere) in the space form Lorentzian space form N_b^{n+1} of constant curvature b is given by*

$$
f_b(r) = \begin{cases} \sqrt{b}\coth(\sqrt{b}\,r) & \text{if } b > 0 \\ 1/r & \text{if } b = 0 \\ \sqrt{-b}\cot(\sqrt{-b}\,r) & \text{if } b < 0 \text{ and } 0 < r < \pi/\sqrt{-c}. \end{cases} \tag{4.4}
$$

Remark 4.4. Note that the future-directed timelike unit normal field globally defined on $\Sigma_b(r)$ is the gradient $-\nabla^N d_p$.

We present now the following comparison results for the Hessian of the Lorentzian distance function when the sectional curvatures of timelike planes of the ambient spacetime are bounded from above or from below, as we did in Proposition 3.8 in the Riemannian case.

Proposition 4.5 ([AHP]). *Let N^{n+1} be an $(n+1)$-dimensional spacetime such that $K_N(\Pi) \leq b$, $b \in \mathbb{R}$, for all timelike planes in N. Assume that there exists a point $p \in N$ such that $\mathcal{I}^+(p) \neq \emptyset$, and let $q \in \mathcal{I}^+(p)$, (with $d_p(q) < \pi/\sqrt{-b}$ when $b < 0$). Then for every spacelike vector $x \in T_qN$ orthogonal to $\nabla^N d_p(q)$ it holds that*

$$
\operatorname{Hess}^N d_p(x,x) \geq -f_b(d_p(q))\langle x, x \rangle, \tag{4.5}
$$

where Hess^N stands for the Hessian operator on N. When $b < 0$ but $d_p(q) \geq \pi/\sqrt{-b}$, then it still holds that

$$
\operatorname{Hess}^N d_p(x,x) \geq -\frac{1}{d_p(q)}\langle x, x \rangle \geq -\frac{\sqrt{-b}}{\pi}\langle x, x \rangle. \tag{4.6}
$$

Proof. The proof follows the ideas of the proof of [EGK, Theorem 3.1]. Let us consider $v = \exp_p^{-1}(q) \in \operatorname{int}(\tilde{\mathcal{I}}^+(p))$ and let $\gamma(t) = \exp_p(tv)$, $0 \leq t < s_p(v)$, the radial future directed unit timelike geodesic with $\gamma(0) = p$ and $\gamma(s) = q$, where $s = d_p(q)$.

As $\gamma'(s) = -\nabla^N d_p(q)$, we have from [EGK, Proposition 3.3], that

$$
\operatorname{Hess}^N d_p(x,x) = -\int_0^s (\langle J'(t), J'(t) \rangle - \langle R(J(t), \gamma'(t))\gamma'(t), J(t) \rangle)dt = I_\gamma(J, J)
$$

where J is the (unique) Jacobi field along γ such that $J(0) = 0$ and $J(s) = x$.

We also know that there are no conjugate points of $\gamma(0)$ along the geodesic γ, because $\gamma : [0, s] \to \mathcal{I}^+(p)$ and $\exp_p : \operatorname{int}(\tilde{\mathcal{I}}^+(p)) \to \mathcal{I}^+(p)$ is a diffeomorphism.

Therefore, using the maximality of the index of Jacobi fields (see [BEE, Theorem 10.23]) we get that

$$
\operatorname{Hess}^N d_p(x,x) = I_\gamma(J, J) \geq I_\gamma(X, X), \tag{4.7}
$$

for every vector field X along γ such that $X(0) = J(0) = 0$, $X(s) = J(s) = x$ and $X(t) \perp \gamma'(t)$ for every t.

The result follows by using the expression of $I_\gamma(X, X)$ in terms of $K(t)$ and the curvature bound $K(t) \leq b$ as it can be found in [AHP]. $\square$

Proposition 4.6 ([AHP]). *Let N^{n+1} be an $(n+1)$-dimensional spacetime such that $K_N(\Pi) \geq b$, $b \in \mathbb{R}$, for all timelike planes in N. Assume that there exists a point $p \in N$ such that $\mathcal{I}^+(p) \neq \emptyset$, and let $q \in \mathcal{I}^+(p)$ (with $d_p(q) < \pi/\sqrt{-b}$ when $b < 0$). Then, for every spacelike vector $x \in T_q N$ orthogonal to $\nabla^N d_p(q)$ it holds that*

$$\mathrm{Hess}^N d_p(x, x) \leq -f_b(d_p(q))\langle x, x \rangle .$$

Combining the previous results, we now have the following extrinsic comparison results for the Hessian of the restricted Lorentzian distance, in the style of Proposition 3.10, when we consider $\psi : \Sigma^n \to N^{n+1}$ a spacelike hypersurface immersed into the spacetime N:

Proposition 4.7 ([AHP]). *Let N^{n+1} be a spacetime such that $K_N(\Pi) \leq b$ for all timelike planes in N. Assume that there exists a point $p \in N$ such that $\mathcal{I}^+(p) \neq \emptyset$, and let $\psi : \Sigma^n \to N^{n+1}$ be a spacelike hypersurface such that $\psi(\Sigma) \subset \mathcal{I}^+(p)$. Let u denote the Lorentzian distance function from p along the hypersurface Σ, (with $u < \pi/\sqrt{-b}$ on Σ when $b < 0$). Then*

$$\mathrm{Hess}^\Sigma u(X, X) \geq -f_b(u)(1 + \langle X, \nabla^\Sigma u \rangle^2) - \langle AX, X \rangle \sqrt{1 + |\nabla^\Sigma u|^2} \qquad (4.8)$$

for every unit tangent vector $X \in T\Sigma$, and

$$\Delta^\Sigma u \geq -f_c(u)(n + |\nabla^\Sigma u|^2) + nH\sqrt{1 + |\nabla^\Sigma u|^2} \quad , \qquad (4.9)$$

where H is the future mean curvature of Σ.

On the other hand, if we assume that $K_N(\Pi) \geq b$ for all timelike planes in N, then the same analysis yields the following – using now Proposition 4.6 instead of Proposition 4.5:

Proposition 4.8. *Let N^{n+1} be a spacetime such that $K_N(\Pi) \geq b$ for all timelike planes in N. Assume that there exists a point $p \in N$ such that $\mathcal{I}^+(p) \neq \emptyset$, and let $\psi : \Sigma^n \to N^{n+1}$ be a spacelike hypersurface such that $\psi(\Sigma) \subset \mathcal{I}^+(p)$. Let u denote the Lorentzian distance function from p along the hypersurface Σ, (with $u < \pi/\sqrt{-b}$ on Σ when $b < 0$). Then*

$$\mathrm{Hess}^\Sigma u(X, X) \leq -f_b(u)(1 + \langle X, \nabla^\Sigma u \rangle^2) - \langle AX, X \rangle \sqrt{1 + |\nabla^\Sigma u|^2} \qquad (4.10)$$

for every unit tangent vector $X \in T\Sigma$, and

$$\Delta^\Sigma u \leq -f_b(u)(n + |\nabla^\Sigma u|^2) + nH\sqrt{1 + |\nabla^\Sigma u|^2} , \qquad (4.11)$$

where H is the future mean curvature of Σ.

5. Concerning the Riemannian Setting and Notation

Returning now to the Riemannian case: Although we indeed do have the possibility of considering 4 basically different settings determined by the choice of p or V as the 'base' of our normal domain and the choice of $K_N \leq b$ or $K_N \geq b$ as the curvature assumption for the ambient space N, we will, however, mainly consider the 'first' of these. Specifically we will (unless otherwise explicitly stated) apply the following assumptions and denotations:

Definition 5.1. A *standard situation* encompasses the following:

(1) P^m denotes an m-dimensional complete minimally immersed submanifold of the Riemannian manifold N^n. We always assume that P has dimension $m \geq 2$.

(2) The sectional curvatures of N are assumed to satisfy $K_N \leq b$, $b \in \mathbb{R}$, cf. Proposition 3.10, equation (3.13).

(3) The intersection of P with a regular ball $B_R(p)$ centered at $p \in P$ (cf. Definition 3.4) is denoted by

$$D_R = D_R(p) = P^m \cap B_R(p),$$

and this is called a minimal extrinsic R-ball of P in N, see the Figures 3–7 of extrinsic balls, which are cut out from some of the well-known minimal surfaces in $\mathbb{R}^3$.

(4) The totally geodesic m-dimensional regular R-ball centered at $\tilde{p}$ in $\mathbb{K}^n(b)$ is denoted by

$$B_R^{b,m} = B_R^{b,m}(\tilde{p}),$$

whose boundary is the $(m-1)$-dimensional sphere

$$\partial B_R^{b,m} = S_R^{b,m-1}.$$

(5) For any given smooth function F of one real variable we denote

$$W_F(r) = F''(r) - F'(r)h_b(r) \text{ for } 0 \leq r \leq R.$$

We may now collect the basic inequalities from our previous analysis as follows.

FIGURE 2. A standard view of the Helicoid with center axis.

FIGURE 3. An extrinsic minimal ball D_R of a Helicoid.

FIGURE 4. Yet another extrinsic minimal ball of a Helicoid.

Proposition 5.2. *In a standard situation with*

$$\begin{cases} F'(r) \geq 0 & and \\ W_F(r) \leq 0 & for \ 0 \leq r \leq R \end{cases} \tag{5.1}$$

we get the following comparison result at all points of the minimal extrinsic R-ball D_R:

$$\begin{aligned} \Delta^P(F \circ r) &\geq W_F(r)|\nabla^P r|^2 + mF'(r)h_b(r) \\ &\geq F''(r) + (m-1)F'(r)h_b(r) \\ &= \Delta^{\mathbb{K}^m(b)}(F \circ r) \,. \end{aligned} \tag{5.2}$$

Proposition 5.3. *In a standard situation with*

$$\begin{cases} F'(r) \leq 0 & and \\ W_F(r) \geq 0 & for \ 0 \leq r \leq R \end{cases} \tag{5.3}$$

we get the following comparison result at all points of the minimal extrinsic R-ball D_R:

$$\begin{aligned} \Delta^P(F \circ r) &\leq W_F(r)|\nabla^P r|^2 + mF'(r)h_b(r) \\ &\leq F''(r) + (m-1)F'(r)h_b(r) \\ &= \Delta^{\mathbb{K}^m(b)}(F \circ r) \,. \end{aligned} \tag{5.4}$$

In the following sections we will find several applications of these basic inequalities for standard situations.

In particular we note the consequences of having equality in either of the equations (5.2) or (5.4):

FIGURE 5. An extrinsic minimal ball of a Catenoid. (Note that the center of the ball is positioned on the "waist" circle of the Catenoid.)

FIGURE 6. An extrinsic minimal ball of Costa's minimal surface.

FIGURE 7. An extrinsic minimal ball of Scherk's surface.

Definition 5.4. The *standard rigidity conclusion* for a standard situation amounts to the following:

(1) D_R is a minimal radial cone from p in N:

$$D_R = D_R(p) = \exp_p\left(\{\, v \in T_pP \subseteq T_pN \mid |v| \leq R \,\}\right) .$$

(2) If $N = \mathbb{K}^n(b)$ then D_R and in fact all of P^m is a totally geodesic submanifold of $\mathbb{K}^n(b)$.

Proposition 5.5. *If a standard situation satisfies that $W_F(r)$ does not vanish for $0 < r \leq R$, and if both inequalities in (5.2) or in (5.4) are equalities for all $r \leq R$, then the standard rigidity conclusions hold.*

Proof. Under the hypotheses we have that

$$|\nabla^P r| = 1 \text{ at every point in } D_R, \text{ so that}$$

$$\nabla^N r = \nabla^P r \text{ at every point },$$

$$(5.5)$$

which implies that the tangent vectors to the radial geodesics from p to points in $D_R(p)$ are contained in TD_R. Hence the geodesics themselves are contained in $D_R(p)$, which must therefore be a cone with top point p. Radial minimal cones in spaces $\mathbb{K}^n(b)$ of constant curvature are totally geodesic, so by analytic continuation all of P^m is totally geodesic in $\mathbb{K}^n(b)$. $\square$

Remark 5.6. The interesting problem of obtaining rigidity conclusions also in cases where $W_F(r)$ vanishes identically for all $r \leq R$ will be addressed briefly in Section 8.

6. Green's Formulae and the Co-area Formula

Now we recall the coarea formula. We follow the lines of [Sa] Chapter II, Section 5. Let (M, g) denote a Riemannian manifold and Ω a precompact domain in M. Let $\psi : \Omega \to \mathbb{R}$ be a smooth function such that $\psi(\Omega) = [a, b]$ with $a < b$. Denote by Ω_0 the set of critical points of ψ. By Sard's theorem, the set of critical values $S_\psi = \psi(\Omega_0)$ has null measure, and the set of regular values $R_\psi = [a, b] - S_\psi$ is open. In particular, for any $t \in R_\psi = [a, b] - S_\psi$, the set $\Gamma(t) := \psi^{-1}(t)$ is a smooth embedded hypersurface in Ω with $\partial\Gamma(t) = \emptyset$. Since $\Gamma(t) \subseteq \Omega - \Omega_0$ then $\nabla\psi$ does not vanish along $\Gamma(t)$; indeed, a unit normal along $\Gamma(t)$ is given by $\nabla\psi/|\nabla\psi|$.

Now we let

$$A(t) = \text{Vol}(\Gamma(t))$$

$$\Omega(t) = \{x \in \overline{\Omega} \mid \psi(x) < t\} \qquad (6.1)$$

$$V(t) = \text{Vol}(\Omega(t)) .$$

Theorem 6.1.

   i) *For every integrable function u on $\overline{\Omega}$:*

$$\int_\Omega u \cdot |\nabla\psi| \, dV = \int_a^b \left(\int_{\Gamma(t)} u \, dA_t \right) dt , \qquad (6.2)$$

   where dA_t is the Riemannian volume element defined from the induced metric g_t on $\Gamma(t)$ from g.

   ii) *The function $V(t)$ is a smooth function on the regular values of ψ given by:*

$$V(t) = \text{Vol}(\Omega_0 \cap \Omega(t)) + \int_a^t \left(\int_{\Gamma(t)} |\nabla\psi|^{-1} \, dA_t \right) ,$$

   and its derivative is

$$\frac{d}{dt} V(t) = \int_{\Gamma(t)} |\nabla\psi|^{-1} \, dA_t . \qquad (6.3)$$

Remark 6.2.

  i) The term $\mathrm{Vol}(\Omega_0 \cap \Omega(t))$ vanishes if Ω_0 has null measure.

  ii) If we set $\Omega(t) = \{x \in \overline{\Omega} \mid \psi(x) > t\}$, then the first assertion of the theorem holds as it stands and the derivative of $V(t) = \mathrm{Vol}(\Omega(t))$ becomes

$$\frac{d}{dt} V(t) = -\int_{\Gamma(t)} |\nabla \psi|^{-1} dA_t .\tag{6.4}$$

  iii) If we set $u = 1$, then equation (6.2) becomes:

$$\int_\Omega |\nabla \psi| \, dV = \int_a^b A(t) \, dt .\tag{6.5}$$

  iv) If Ω_0 does not have null measure, then the function $V(t)$ need not be C^1 on (a, b). Indeed, it can be discontinuous in some $t_0 \in S_\psi$.

  v) Theorem 6.1 holds if we consider a continuous function $\psi : \overline{\Omega} \to \mathbb{R}$ which is smooth in Ω and vanishes on the boundary of Ω.

We will also make frequent use of the divergence theorem and Green's formulae. We therefore state them here for further reference.

Theorem 6.3. *Let X denote a smooth vector field on M with compact support on $\overline{\Omega}$. Then*

$$\int_\Omega \mathrm{div}(X) \, dV = \int_{\partial \Omega} \langle X , \nu \rangle \, dA ,\tag{6.6}$$

where ν is the outward (from Ω) pointing unit normal vector field along $\partial \Omega$.

Theorem 6.4. *Let h and f denote smooth functions on M such that both h, f, and the vector field $h \nabla f$ have compact support in $\overline{\Omega}$. Then*

$$\int_\Omega (h \Delta f + \langle \nabla h, \nabla f \rangle \,) \, dV = \int_{\partial \Omega} h \cdot \nu(f) \, dA \quad and \tag{6.7}$$

$$\int_\Omega (h \Delta f - f \Delta h) \, dV = \int_{\partial \Omega} (h \cdot \nu(f) - f \cdot \nu(h)) \, dA .\tag{6.8}$$

7. The First Dirichlet Eigenvalue Comparison Theorem

Following standard notations and setting (see, e.g., [Cha1] or in this context the seminal survey by Grigoryan in [Gri1]), for any precompact open set Ω in a Riemannian manifold M we denote by $\lambda(\Omega)$ the smallest number λ for which the following Dirichlet eigenvalue problem has a non-zero solution

$$\begin{cases} \Delta u + \lambda u & = 0 \text{ at all points } x \text{ in } \Omega \\ \quad\quad u(x) = 0 \text{ at all points } x \text{ in } \partial \Omega . \end{cases}\tag{7.1}$$

We shall need the following beautiful observation due to Barta:

Theorem 7.1 ([B], [Cha1]). *Consider any smooth function f on a domain Ω which satisfies $f_{|\Omega} > 0$ and $f_{|\partial\Omega} = 0$, and let $\lambda(\Omega)$ denote the first eigenvalue of the Dirichlet problem for Ω. Then*

$$\inf_{\Omega}\left(\frac{\Delta f}{f}\right) \leq -\lambda(\Omega) \leq \sup_{\Omega}\left(\frac{\Delta f}{f}\right). \tag{7.2}$$

If equality occurs in one of the inequalities, then they are both equalities, and f is an eigenfunction for Ω corresponding to the eigenvalue $\lambda(\Omega)$.

Proof. Let ϕ be an eigenfunction for Ω corresponding to $\lambda(\Omega)$. Then $\phi_{|\Omega} > 0$ and $\phi_{|\partial\Omega} = 0$. If we let h denote the difference $h = \phi - f$, then

$$-\lambda(\Omega) = \frac{\Delta\phi}{\phi} = \frac{\Delta f}{f} + \frac{f\Delta h - h\Delta f}{f(f+h)}$$

$$= \inf_{\Omega}\left(\frac{\Delta f}{f}\right) + \sup_{\Omega}\left(\frac{f\Delta h - h\Delta f}{f(f+h)}\right) \tag{7.3}$$

$$= \sup_{\Omega}\left(\frac{\Delta f}{f}\right) + \inf_{\Omega}\left(\frac{f\Delta h - h\Delta f}{f(f+h)}\right).$$

Here the supremum, $\sup_{\Omega}\left(\dfrac{f\Delta h - h\Delta f}{f(f+h)}\right)$ is necessarily positive since

$$f(f+h)|_{\Omega} > 0, \tag{7.4}$$

and since by Green's second formula (6.8) in Theorem 6.4 we have

$$\int_{\Omega}(f\Delta h - h\Delta f)\, dV = 0. \tag{7.5}$$

For the same reason, the infimum, $\inf_{\Omega}\left(\dfrac{f\Delta h - h\Delta f}{f(f+h)}\right)$ is necessarily negative. This gives the first part of the theorem. If equality occurs, then $(f\Delta h - h\Delta f)$ must vanish identically on Ω, so that $-\lambda(\Omega) = \dfrac{\Delta f}{f}$, which gives the last part of the statement. □

As already alluded to in the introduction, the key heuristic message of this report is that the Laplacian is a particularly 'swift actor' on minimal submanifolds (i.e., minimal extrinsic regular R-balls D_R) in ambient spaces with an upper bound b on its sectional curvatures. This is to be understood in comparison with the 'action' of the Laplacian on totally geodesic R-balls $B_R^{b,m}$ in spaces of constant curvature b. In this section we will use Barta's theorem to show that this phenomenon can indeed be 'heard' by 'listening' to the bass note of the Dirichlet spectrum of any given D_R.

Theorem 7.2. *The respective first Dirichlet eigenvalues satisfy*

$$\lambda(D_R) \geq \lambda(B_R^{b,m}). \tag{7.6}$$

Equality is attained if and only if the standard rigidity conclusions hold.

Proof. In view of Barta's theorem we only have to construct a smooth positive function T on D_R, such that $T_{|\partial D_R} = 0$ and such that

$$\Delta^P T \le -\tilde{\lambda}T \,, \tag{7.7}$$

where we have denoted the first eigenvalue $\lambda(B_R^{b,m})$ by $\tilde{\lambda}$.

In our setting there is but one natural candidate for the function T, namely the transplanted first Dirichlet eigenfunction T from the constant curvature totally geodesic ball $B_R^{b,m}$. Let us recall that T is then a radially symmetric function $T(r)$ which satisfies the differential equation

$$T''(r) + (m-1)h_b(r)T'(r) = -\tilde{\lambda}T(r) \,, \tag{7.8}$$

and the boundary conditions $T'(0) = 0$ and $T(R) = 0$.

The solution $T(r)$ satisfies

Lemma 7.3.

$$T'(r) < 0 \,, \text{ for all } r \in \,]0, R] \,, \text{ and}$$
$$T''(r) - h_b(r)T'(r) > 0 \,, \text{ for all } r \in \,]0, R] \,\,. \tag{7.9}$$

Proof of lemma. The first claim follows from the positivity of the first eigenfunction. Indeed, suppose that $T'(r_0) = 0$ for some r_0. Then at this value of r we have $T''(r_0) = -\tilde{\lambda}T(r_0) < 0$ so that $T'(r)$ cannot grow positive from the zero value at r_0.

The second claim follows from reorganizing the equation (7.8) to

$$T''(r) - h_b(r)T'(r) = -mT'(r)h_b(r) - \tilde{\lambda}T(r) \,. \tag{7.10}$$

In accordance with previous notation let us denote the left-hand side of this equation (and hence the right-hand side as well) by $W_T(r)$. We must then show, that $W_T(r) \ge 0$ for all $r \in \,]0, R]$. Firstly $W_T(0) = 0$. Indeed, for the rotationally symmetric metric around $\tilde{p}$ we have

$$\Delta^{\mathbb{K}^m(b)}T(r)_{|r=0} = mT''(r)_{|r=0} \,, \tag{7.11}$$

so that in our setting

$$mT''(r)_{|r=0} = -\tilde{\lambda}T(0) \,. \tag{7.12}$$

An application of L'Hospital's rule then gives

$$\lim_{r\to 0}\left(mT'(r)h_b(r) + \tilde{\lambda}T(0)\right) = W_T(0) = 0 \,. \tag{7.13}$$

Furthermore,

$$\begin{aligned}
W_T'(r) &= -mT''(r)h_b(r) - mT'(r)h_b'(r) - \tilde{\lambda}T'(r) \\
&= -mh_b(r)\left(T''(r) - T'(r)h_b(r)\right) + (mb - \tilde{\lambda})T'(r) \tag{7.14} \\
&= -mh_b(r)W_T(r) + (mb - \tilde{\lambda})T'(r) \,.
\end{aligned}$$

The factor $(mb - \tilde{\lambda})$ is always negative; For positive b (which is the only case to consider) we have from the domain monotonicity of the Dirichlet spectrum:

$$\tilde{\lambda} = \lambda(B_R^{b,m}) > \lambda(B_{\frac{\pi}{2\sqrt{b}}}^{b,m}) = mb \,. \tag{7.15}$$

Following [CheLY2] pp. 1042–1043, we now prove that

$$W_T''(0) > 0 \,. \tag{7.16}$$

For transparency we only do the calculations for the case $b = 0$, where $h_b(r) = 1/r$. The general cases follow almost verbatim. Then we have for $W = W_T = W_T(r)$:

$$W'' = \tilde{\lambda} T'' - \frac{m}{r}\left[T''' - \frac{2T''}{r} + \frac{2T'}{r^2} \right] \,, \tag{7.17}$$

so that

$$W''(0) = \frac{\tilde{\lambda}^2}{m} T(0) - m \lim_{r\to 0}\left(\frac{1}{r}\left[T''' - \frac{2T''}{r} + \frac{2T'}{r^2} \right] \right) \,. \tag{7.18}$$

Again according to L'Hospital's rule, the latter limit is just the limit of the derivative of the square bracket, so that

$$\lim_{r\to 0}\left(\frac{1}{r}\left[T''' - \frac{2T''}{r} + \frac{2T'}{r^2} \right] \right)$$
$$= \lim_{r\to 0}\left[T'''' - \frac{2T'''}{r} + \frac{4T''}{r^2} - \frac{4T'}{r^3} \right] \,. \tag{7.19}$$

If we differentiate (7.8) twice (still using $h_b(r) = 1/r$), we get

$$T'''' + \frac{m-1}{r}\left[T''' - \frac{2T''}{r} + \frac{2T'}{r^2} \right] = -\tilde{\lambda} T'' \,, \tag{7.20}$$

so that (7.19) gives

$$\lim_{r\to 0}\left(\frac{1}{r}\left[T''' - \frac{2T''}{r} + \frac{2T'}{r^2} \right] \right)$$
$$= \lim_{r\to 0}\left[-\tilde{\lambda} T'' - \frac{m+1}{r}\left(T''' - \frac{2T''}{r} + \frac{2T'}{r^2} \right) \right] \tag{7.21}$$
$$= \frac{\tilde{\lambda}^2}{m} T(0) - (m+1) \lim_{r\to 0}\left(\frac{1}{r}\left[T''' - \frac{2T''}{r} + \frac{2T'}{r^2} \right] \right) \,.$$

Therefore

$$(m+2) \lim_{r\to 0}\left(\frac{1}{r}\left[T''' - \frac{2T''}{r} + \frac{2T'}{r^2} \right] \right) = \frac{\tilde{\lambda}^2}{m} T(0) \,, \tag{7.22}$$

which in combination with (7.18) gives the desired inequality:

$$W''(0) = \frac{\tilde{\lambda}^2}{m} T(0) - \frac{\tilde{\lambda}^2}{m+2} T(0)$$
$$= \frac{2\tilde{\lambda}^2}{m(m+2)} T(0) > 0 \,. \tag{7.23}$$

From this we conclude that the Taylor expansion of $W_T(r)$ 'lifts' the function positively away from 0 for sufficiently small positive values of r.

Finally if we assume then for contradiction, that for some smallest positive r_0 we have $W_T(r_0) = 0$, then (7.14) shows that $W_T'(r_0) > 0$, which is the desired contradiction. This proves the lemma. □

We now continue the proof of Theorem 7.2. According to Proposition 3.13 the Laplacian of the transplanted function $T(r)$ on D_R satisfies:

$$\begin{aligned}
\Delta^P T(r) &\leq W_T(r)|\nabla^P(r)|^2 + mT'(r)h_b(r) \\
&\leq W_T(r) + mT'(r)h_b(r) \\
&= \Delta^{\mathbb{K}^m(b)} T(r) \\
&= -\tilde{\lambda} T(r) ,
\end{aligned} \tag{7.24}$$

which gives the desired inequality (7.7). Equality is attained there if and only if

$$|\nabla^P(r)| = 1 \text{ for all } r \leq R , \tag{7.25}$$

which means precisely that D_R is a radial geodesic cone centered at p. This finishes the proof of Theorem 7.2. □

Remark 7.4. The domain monotonicity property of the Dirichlet spectrum implies that $\lambda(\Omega) \geq \lambda(D_R) \geq \lambda(B_R^{b,m})$ for any subdomain $\Omega \subseteq D_R$, and here equality is attained in the first inequality if and only if $\Omega = D_R$.

Theorem 7.2 was originally obtained in [Ma3] and was found there as a corollary from a more general comparison theorem concerning the Dirichlet heat kernels of D_R and $B_R^{b,m}$ respectively. The heat kernel comparison technique was developed by Cheng, Li and Yau for constant curvature ambient spaces in [CheLY2], see also [CheLY1].

For completeness – and also for later reference – let us state these results here, but only briefly and without proofs.

The p-centered Dirichlet heat kernel on a given domain Ω is the function $H(p, x, t)$ satisfying the heat equation and boundary conditions:

$$\begin{cases}
\left(\Delta - \dfrac{\partial}{\partial t}\right) H(p, x, t) = 0 \text{ for all } x \in \Omega - \{p\} \\
H(p, x, 0) = \delta_p(x) \text{ (Dirac's delta, based at } p) \\
H(p, x, t)_{|\partial\Omega} = 0 \text{ for all } t > 0 .
\end{cases} \tag{7.26}$$

In the totally geodesic R-ball $B_R^{b,m}(\tilde{p})$ in $\mathbb{K}^n(b)$ the $\tilde{p}$-centered heat kernel $\tilde{H}$ is radially symmetric:

$$\tilde{H}(\tilde{p}, x, t) = \tilde{H}(\text{dist}_{\mathbb{K}^n(b)}(\tilde{p}, x), t) , \tag{7.27}$$

so that we may transplant $\tilde{H}$ to D_R via the distance function from p in N and compare the heat kernels pointwise in D_R:

Theorem 7.5 ([Ma3]). *Suppose the domain Ω is contained in the extrinsic minimal R-ball $D_R(p)$ in a minimal submanifold P^m in N which has the upper bound b on its sectional curvatures K_N. Then*

$$H(p, x, t) \leq \tilde{H}(r(x), t) \text{ for all } x \in \Omega \text{ and for all } t \in \mathbb{R}_+ , \tag{7.28}$$

where as usual $r(x)$ denotes the distance in N from p to x.

Again this supports the heuristic idea, that the Laplace operator moves the heat to the boundary quicker when the domain is minimal and when the ambient space has smaller curvature in comparison with the constant curvature setting.

As already mentioned, the first Dirichlet eigenvalue comparison result $\lambda_1 = \lambda(\Omega) \geq \lambda(B_R^{b,m}) = \tilde{\lambda}_1$ is in fact a direct consequence of the heat kernel comparison result. This may be seen as follows:

In particular we have that

$$H(p,p,t) \leq \tilde{H}(0,t) \text{ for all } t > 0 , \tag{7.29}$$

which in terms of the respective Dirichlet eigenfunction expansions reads as follows, see [Cha1] p. 13:

$$\sum_{i=1}^{\infty} e^{-\lambda_i t} \phi_i(p)^2 \leq \sum_{i=1}^{\infty} e^{-\tilde{\lambda}_i t} \tilde{\phi}_i(p)^2 \text{ for all } t > 0 , \tag{7.30}$$

and hence for all $t > 0$ and for all μ

$$\sum_{i=1}^{\infty} e^{(\mu-\lambda_i)t} \phi_i(p)^2 \leq \sum_{i=1}^{\infty} e^{(\mu-\tilde{\lambda}_i)t} \tilde{\phi}_i(p)^2 . \tag{7.31}$$

In particular we observe, that both sides of this latter inequality must go to ∞ for $t \to \infty$ when $\mu > \lambda_1$. Now assume for contradiction that $\lambda_1 < \tilde{\lambda}_1$. Then put $\mu = \lambda_1 + \epsilon < \tilde{\lambda}_1$. For this choice of μ we now have that the right-hand side of (7.31) goes to 0 whereas the left-hand side still goes to ∞, which is the desired contradiction.

At this point we may also note that if $\lambda(D_R)$ stays positively bounded away from 0 when $R \to \infty$ (assuming that $i(N) = \infty$), then the t-integral of the p-centered heat kernels are bounded, and this implies in turn the existence of a bounded Green function on P, so that P has finite resistance to infinity. Indeed, in analogy with the above considerations we can extract the first eigenvalue of P from the heat kernel as follows (see, e.g., [ChaK]):

$$\lim_{t \to \infty} \left(\frac{\ln H(p,x,t)}{t} \right) = -\lambda(P) . \tag{7.32}$$

In consequence, if $\lambda(P) > 0$, then the heat kernel decays exponentially so that the following Green function integral converges for all $x \neq p$

$$G(p,x) = \int_0^{\infty} H(p,x,t) \, dt . \tag{7.33}$$

For $b < 0$ we have from Theorem 7.2 and from, e.g., [Cha1] p. 46:

$$\lambda(P) = \lim_{R \to \infty} \lambda(D_R) \geq \lim_{R \to \infty} \lambda(B_R^{b,m}) = (m-1)^2(-b)/4 > 0 . \tag{7.34}$$

In view of the Kelvin–Nevanlinna–Royden criteria for hyperbolicity (which will be presented in Section 15), we have thus established at least a partial idea or hint in the direction of the following theorem, which we will outline and study further – in a precise sense, even quantitatively – in Section 14.

Theorem 7.6 ([MaP4]). *Let P^m denote a minimally immersed submanifold of dimension $m \geq 2$ in a Cartan Hadamard manifold N^n with curvatures $K_N \leq b \leq 0$. If $b = 0$ we assume further that $m \geq 3$. Then P is hyperbolic in the sense that it has finite effective resistance to infinity.*

Remark 7.7. We observe that the case $b = 0$, $m \geq 3$ does not follow from the eigenvalue comparison, but is a consequence of the capacity comparison theorem, that we obtain in Section 14.

8. Isoperimetric Relations

In this and the following two sections we survey some comparison results concerning inequalities of isoperimetric type, mean exit times and capacities, respectively, for extrinsic minimal balls in ambient spaces with an upper bound on sectional curvature. This has been developed in a series of papers, see [Pa] and [MaP1]–[MaP4].

We will still assume a standard situation as in the previous section, i.e., D_R denotes an extrinsic minimal ball of a minimal submanifold P in an ambient space N with the upper bound b on the sectional curvatures.

Proposition 8.1. *We define the following function of $t \in \mathbb{R}_+ \cup \{0\}$ for every $b \in \mathbb{R}$, for every $q \in \mathbb{R}$, and for every dimension $m \geq 2$:*

$$L_q^{b,m}(t) = q\left(\frac{\mathrm{Vol}(S_t^{b,m-1})}{mh_b(t)} - \mathrm{Vol}(B_t^{b,m})\right). \tag{8.1}$$

Then

$$L_q^{b,m}(0) = 0 \text{ for all } b, q, \text{ and } m, \tag{8.2}$$

and

$$\mathrm{sign}(\frac{d}{dt}L_q^{b,m}(t)) = \mathrm{sign}(b\,q) \text{ for all } b, q, m, \text{ and } t > 0. \tag{8.3}$$

Proof. This follows from a direct computation using the definition of $h_b(t)$ from equation (3.5) together with the volume formulae (cf. [Gr])

$$\mathrm{Vol}(B_t^{b,m}) = \mathrm{Vol}(S_1^{0,m-1}) \cdot \int_0^t (Q_b(u))^{m-1}\, du \tag{8.4}$$

$$\mathrm{Vol}(S_t^{b,m-1}) = \mathrm{Vol}(S_1^{0,m-1}) \cdot (Q_b(t))^{m-1}. \qquad \square$$

9. A Consequence of the Co-area Formula

The co-area equation (6.4) applied to our setting gives the following

Proposition 9.1. *Let $D_R(p)$ denote a regular extrinsic minimal ball of P with center p in N. Then*

$$\frac{d}{du}\,\mathrm{Vol}(D_u) \geq \mathrm{Vol}(\partial D_u) \text{ for all } u \leq R. \tag{9.1}$$

Proof. We let $f : \overline{D}_R \to \mathbb{R}$ denote the function $f(x) = R - r(x)$, which clearly vanishes on the boundary of D_R and is smooth except at p. Following the notation of the co-area formula we further let

$$\Omega(t) = D_{(R-t)}$$
$$V(t) = \mathrm{Vol}(D_{(R-t)}) \text{ and} \tag{9.2}$$
$$\Sigma(t) = \partial D_{(R-t)} .$$

Then

$$\mathrm{Vol}(D_u) = V(R-u) \text{ so that}$$
$$\frac{d}{du}\mathrm{Vol}(D_u) = -V'(t)\big|_{t=R-u} . \tag{9.3}$$

The co-area equation (6.4) now gives

$$-V'(t) = \int_{\partial D_{(R-t)}} |\nabla^P r|^{-1} \, dA$$
$$\geq \mathrm{Vol}(\partial D_{(R-t)}) \tag{9.4}$$
$$= \mathrm{Vol}(\partial D_u) ,$$

and this proves the statement.   □

Exercise 9.2. Explain why the non-smoothness of the function f at p does not create problems for the application of equation (6.4) in this proof although smoothness is one of the assumptions in Theorem 6.1.

10. The Fundamental Differential Equation

As already mentioned we are particularly interested in the general solutions to the following ordinary differential equation, where q denotes a constant (usually $q \leq 0$)

$$\psi''(r) + (m-1)\psi'(r)h_b(r) = q , \tag{10.1}$$

or, equivalently

$$\Gamma'(r) + (m-1)\Gamma(r)h_b(r) = q , \tag{10.2}$$

where we have defined the auxiliary function $\Gamma(r) = \psi'(r)$.

Equation (10.1) is the 'radial' version of the Poisson (Laplace) equation

$$\Delta^{\mathbb{K}^m(b)} = q . \tag{10.3}$$

In order to obtain the comparison relations for the solutions of the corresponding equation on P, we need effective inequalities for the Laplacian $\Delta^P \psi(r(x))$ of the transplanted function $\psi(r(x)) = \psi(\mathrm{dist}_N(p,x))$ on the minimal submanifold P within the regular extrinsic ball D_R. As we know this needs good control over the sectional curvatures K_N of N and also over the signs of the intermediary expressions $\Gamma(r)$ and $\Gamma'(r) - \Gamma(r)h_b(r)$. Therefore we now display the general solutions to equation (10.2) with special emphasis on the values of these signs.

Proposition 10.1. *The general solutions to the equation* (10.2) *(without imposing any boundary conditions) are given by*

$$\Gamma(r) = Q_b(r)^{1-m} \left(C + q \int_0^r Q_b(t)^{m-1} dt \right)$$

$$= \text{Vol}(S_r^{b,m-1})^{-1} \left(\tilde{C} + q \, \text{Vol}(B_r^{b,m}) \right) ,$$

(10.4)

where C, *and* $\tilde{C}$ *respectively, are arbitrary constants of integration. They are related as follows:* $\tilde{C} = \text{Vol}(S_1^{0,m-1}) \cdot C$. *The sign of* $\Gamma(r)$ *is clearly controlled by these constants. The solutions* $\Gamma(r)$ *satisfy*

$$\Gamma'(r) - \Gamma(r)h_b(r) \geq 0 \;\; \text{if and only if}$$

$$\tilde{C} \leq L_q^{b,m}(r) .$$

(10.5)

In particular we note the following consequences:

Corollary 10.2. *Let* $\tilde{C} \leq L_q^{b,m}(r)$ *and assume that* $q \leq 0$. *Then*

$$\begin{cases} \Gamma'(r) - \Gamma(r)h_b(r) \geq 0 \;\; \text{and} \\ \qquad \Gamma(r) \qquad\quad \leq 0 \;\; . \end{cases}$$

(10.6)

Corollary 10.3. *Let* $\tilde{C} \geq -q \, \text{Vol}(B_r^{b,m})$ *and assume again that* $q \leq 0$. *Then* $\tilde{C} \geq L_q^{b,m}(r)$ *and*

$$\begin{cases} \Gamma'(r) - \Gamma(r)h_b(r) \leq 0 \;\; \text{and} \\ \qquad \Gamma(r) \qquad\quad \geq 0 \;\; . \end{cases}$$

(10.7)

Remark 10.4. Since $\tilde{C}$ is a constant we obviously have to choose it so that the respective hypotheses (concerning $\Gamma(r)$) in our applications of the corresponding radially symmetric solutions will be satisfied for all r in the standard range $[0, R]$. This is precisely what we are going to pinpoint below.

11. Isoperimetric Comparison

In the general notation of the standard situations we have the following results for all regular extrinsic minimal balls D_R of P^m in N.

Theorem 11.1 ([Pa], [MaP1]). *For every* $b \in \mathbb{R}$ *we have*

$$\text{Vol}(D_R) \geq \text{Vol}(B_R^{b,m}) .$$

(11.1)

If equality occurs then the standard rigidity conclusions given in Definition 5.4 hold.

Proof. Let $\Gamma(r)$ denote the function in (10.4) with $\tilde{C} = \text{Vol}(B_R^{b,m})$ and $q = -1$ and let $\psi(r)$ denote any of the corresponding solutions to equation (10.1).

From Corollary 10.3, the general Laplacian comparison result, equation (3.16) of Proposition 3.13, and from Corollary 3.14 we then have

$$\Delta^P \psi(r(x)) \geq -1 . \tag{11.2}$$

The divergence theorem then gives

$$
\begin{aligned}
\text{Vol}(D_R) - \text{Vol}(D_\epsilon) &= \int_{D_R - D_\epsilon} 1 \, dV \\
&\geq -\int_{D_R - D_\epsilon} \Delta^P \psi(r(x)) \, dV \\
&= -\int_{D_R - D_\epsilon} \text{div}(\nabla^P \psi(r(x))) \, dV \tag{11.3} \\
&= -\Gamma(R) \cdot \int_{\partial D_R} |\nabla^P r(x)| \, dA \\
&\quad + \Gamma(\epsilon) \cdot \int_{\partial D_\epsilon} |\nabla^P r(x)| \, dA .
\end{aligned}
$$

Since

$$\lim_{\epsilon \to 0} \left(\Gamma(\epsilon) \cdot \int_{\partial D_\epsilon} |\nabla^P r(x)| \, dA \right) = \lim_{\epsilon \to 0} \left(\Gamma(\epsilon) \cdot \text{Vol}(S_\epsilon^{b,m-1}) \right) = \tilde{C} , \tag{11.4}$$

we obtain

$$\text{Vol}(D_R) \geq -\Gamma(R) \cdot \int_{\partial D_R} |\nabla^P r(x)| \, dA + \tilde{C} = \text{Vol}(B_R^{b,m}) , \tag{11.5}$$

where we have used that for the actual choice of $\tilde{C}$ we have $\Gamma(R) = 0$. This proves the result. Since $\tilde{C} \neq 0$ equality in (11.1) implies that D_R is a radial cone, and the other rigidity conclusions follow. □

Remark 11.2. The inequality (11.1) was also obtained in [Ma3] using similar heat kernel comparison techniques.

Conversely we also obtain upper bounds on the volumes of extrinsic minimal balls as follows:

Theorem 11.3 ([Pa], [MaP1]). *With the standard setting as above we get the following:*
For $b > 0$:

$$
\begin{aligned}
\text{Vol}(D_R) &\leq \text{Vol}(B_R^{b,m}) \\
&\quad + \frac{1}{m \, h_b(R)} \cdot \left(\int_{\partial D_R} |\nabla^P r| \, dA - \text{Vol}(S_R^{b,m-1}) \right) . \tag{11.6}
\end{aligned}
$$

For $b \leq 0$:

$$\text{Vol}(D_R) \leq \frac{\text{Vol}(B_R^{b,m})}{\text{Vol}(S_R^{b,m-1})} \cdot \int_{\partial D_R} |\nabla^P r| \, dA . \tag{11.7}$$

If equality occurs in (11.6) or (11.7), then the standard rigidity conclusions in 5.4 hold – except when $b = 0$ in which case we can only conclude, that D_R is radial at the boundary in the sense that $|\nabla^P r(x)| = 1$ for all $x \in \partial D_R$.

Proof. Again we let $\Gamma(r)$ denote the function in (10.4) but now with $\tilde{C} = L^{b,m}_{-1}(R)$ for $b > 0$ and $\tilde{C} = 0$ for $b \leq 0$. We also choose $q = -1$, and let $\psi(r)$ denote any of the corresponding solutions to equation (10.1). Then the Laplace inequality is reversed:

$$\Delta^P \psi(r(x)) \leq -1 , \tag{11.8}$$

so that similarly to the proof of Theorem 11.1

$$\mathrm{Vol}(D_R) \leq -\Gamma(R) \cdot \int_{\partial D_R} |\nabla^P r(x)| \, dA + \tilde{C} . \tag{11.9}$$

For $b > 0$ and the (best possible) choice $\tilde{C} = L^{b,m}_{-1}(R)$ we get

$$\Gamma(R) = -\frac{1}{m \, h_b(R)} . \tag{11.10}$$

For $b \leq 0$ and the (best possible) choice $\tilde{C} = 0$ we get

$$\Gamma(R) = -\frac{\mathrm{Vol}(B^{b,m}_R)}{\mathrm{Vol}(S^{b,m-1}_R)} . \tag{11.11}$$

Inserting these expressions into (11.9) now gives the statements of the theorem. Except when $b = \tilde{C} = 0$, equality in one of the equations implies that D_R is a radial cone. When $b = \tilde{C} = 0$ we can only conclude from (11.9), that $|\nabla^P r(x)| = 1$ for $x \in \partial D_R$. $\qquad\square$

Corollary 11.4. *For every $b \in \mathbb{R}$ we have*

$$\mathrm{Vol}(S^{b,m-1}_R) \leq \int_{\partial D_R} |\nabla^P r| \, dA \leq \mathrm{Vol}(\partial D_R) . \tag{11.12}$$

If equality is attained in the first inequality, then the standard rigidity conclusions hold.

Proof. The first inequality follows directly from combining Theorems 11.3 and 11.1. $\qquad\square$

We also point out the following direct consequence of (11.7) in Theorem 11.3

Corollary 11.5 ([Pa]). *Suppose that $K_N \leq b$ and that $b \leq 0$. Then*

$$\frac{\mathrm{Vol}(D_R)}{\mathrm{Vol}(\partial D_R)} \leq \frac{\mathrm{Vol}(B^{b,m}_R)}{\mathrm{Vol}(S^{b,m-1}_R)} . \tag{11.13}$$

If equality holds in (11.13) and if $b < 0$, then the standard rigidity conclusions hold.

Corollary 11.6 (See [MaP2]). *Suppose that $K_N \leq b$ and that $b > 0$. Then*

$$\frac{d}{dr}\left(\frac{\mathrm{Vol}(D_r) - \mathrm{Vol}(B_r^{b,m})}{\mathrm{Vol}(S_r^{b,m})}\right) \geq 0. \tag{11.14}$$

If equality is attained for some value of r, say $r = R$, then it holds for all $r \leq R$, and the standard rigidity conclusions hold for D_R.

Proof. From Proposition 9.1 and equation (11.6) we get

$$
\begin{aligned}
\mathrm{Vol}(D_r) &\leq \mathrm{Vol}(B_r^{b,m}) \\
&\quad + \frac{1}{m\,h_b(r)} \cdot \left(\mathrm{Vol}(\partial D_r) - \mathrm{Vol}(S_r^{b,m-1})\right) \\
&\leq \mathrm{Vol}(B_r^{b,m}) \\
&\quad + \frac{1}{m\,h_b(r)} \cdot \frac{d}{dr}\left(\mathrm{Vol}(D_r) - \mathrm{Vol}(B_r^{b,m})\right),
\end{aligned} \tag{11.15}
$$

so that

$$
\begin{aligned}
\frac{d}{dr}\ln\left(\mathrm{Vol}(D_r) - \mathrm{Vol}(B_r^{b,m})\right) &= \frac{\frac{d}{dr}\left(\mathrm{Vol}(D_r) - \mathrm{Vol}(B_r^{b,m})\right)}{\left(\mathrm{Vol}(D_r) - \mathrm{Vol}(B_r^{b,m})\right)} \\
&\geq m\,h_b(r) \\
&= \frac{d}{dr}\ln\left(Q_b(r)^m\right) \\
&= \frac{d}{dr}\ln\mathrm{Vol}(S_r^{b,m}).
\end{aligned} \tag{11.16}
$$

Hence we get

$$\frac{d}{dr}\ln\frac{\left(\mathrm{Vol}(D_r) - \mathrm{Vol}(B_r^{b,m})\right)}{\mathrm{Vol}(S_r^{b,m})} \geq 0, \tag{11.17}$$

and this proves the corollary. □

In relation to Corollary 11.5 we note the specific problem of obtaining rigidity conclusions in case $b = 0$ when assuming the equality:

$$\frac{\mathrm{Vol}(D_R)}{\mathrm{Vol}(\partial D_R)} = \frac{\mathrm{Vol}(B_R^{0,m})}{\mathrm{Vol}(S_R^{0,m-1})}. \tag{11.18}$$

It is an interesting open problem to find the 'minimal' extra conditions under which the rigidity conclusions will hold for this situation.

In [MaP3] we have shown the following rigidity theorem in this direction:

Theorem 11.7. *In a standard situation with $N = \mathbb{R}^n$ we assume the equality (11.18). If P^m is a hypersurface in $\mathbb{R}^n$ we assume $m \geq 3$, and otherwise we*

assume $m > 3$. *Suppose furthermore that the boundary of the regular ball* D_R
satisfies the bounded curvature assumption

$$|\alpha_P|^2 \leq \left(\frac{m-1}{m-2}\right) \cdot R^{-2} \tag{11.19}$$

*together with the following assumption on parallelism involving the radial gradient
field* $\nabla^{\mathbb{R}^n} r$ *of the distance function:*

$$D^P_X \alpha_P \left(\nabla^{\mathbb{R}^n} r, \nabla^{\mathbb{R}^n} r\right) = 0 \tag{11.20}$$

for all tangent vector fields X *to the boundary* ∂D_R. *Then* D_R *and hence all of* P
is a totally geodesic flat submanifold in $\mathbb{R}^n$.

Remark 11.8. An obviously quite relevant observation for the 'flatness' problem
is a result from [DHKW] Vol. I, p. 341 ff., which shows flatness in the case of
extrinsic *simply-connected* minimal two-dimensional balls in $\mathbb{R}^3$ which intersect the
corresponding R-sphere in $\mathbb{R}^3$ orthogonally at the boundary (as indeed we know
is satisfied in our situation under the assumption of equality (11.18) – in view of
Theorem 11.3). The proof uses (naturally) the theory of holomorphic functions
available for two-dimensional minimal surfaces. It is not clear how we may use
simple connectedness or some other topological restriction to obtain a similar
result for higher dimensions and co-dimensions.

12. Mean Exit Times and Moment Spectra

In this and the next Sections 13 and 14 we will apply the techniques developed in
the previous sections to establish similar comparison theorems for mean exit times
(and their so-called moment spectra) and for the effective resistances of extrinsic
annuli $A_{\rho,R}$ (in the setting of so-called weighted manifolds), respectively.

The mean exit time function E on D_R is the solution to the simplest possible
Poisson equation $\Delta^P E(x) = -1$, which has already played a significant, however
only auxiliary, rôle for the establishment of the isoperimetric comparison in the
previous section. Here we encounter the Poisson solution in the following general
probabilistic arena: Suppose Ω is a smoothly bounded domain with compact clo-
sure in a Riemannian manifold M. Let X_t denote Brownian motion in M (with
infinitesimal generator Δ^M). For any given $x \in \Omega$ we denote by $\mathbb{P}^x$ the probability
measure that charges Brownian paths from x. The first exit time from Ω is then

$$\tau = \inf\{t \geq 0 \; : \; X_t \notin \Omega\} . \tag{12.1}$$

The expectation of the value of the first exit time τ with respect to $\mathbb{P}^x$ is then
the mean exit time from x, $E(x) = \mathbb{E}^x(\tau)$. We refer to [Dy] for an early derivation
of the following simple characterization of $E(x)$:

Proposition 12.1. *For a smooth precompact domain* Ω *in a Riemannian manifold*
M, *the mean exit time function* $E : \overline{\Omega} \to \mathbb{R}_+ \cup \{0\}$ *is given by the continuous*

solution to the boundary value problem

$$
\begin{cases}
\Delta^M E(x) = -1 & \text{at all } x \in \Omega, \text{ and} \\
E(x) \quad = \quad 0 & \text{at all } x \in \partial\Omega \ .
\end{cases}
\tag{12.2}
$$

From our previous analysis we know the solutions to this problem for the totally geodesic balls $B_R^{b,m}$ in the space forms $\mathbb{K}^m(b)$ of constant curvature b. Let us denote these solutions by $E_R^{b,m}(r)$, since they are radially symmetric:

Proposition 12.2 (See [MaP1], [Ma6]). *The mean exit time $E_R^{b,m}(r)$ from $B_R^{b,m}(\tilde{p})$ for the Brownian motion from a point of distance r from the center $\tilde{p}$ is given explicitly by*

$$
E_R^{b,m}(r) = \int_r^R \left(\text{Vol}(S_t^{b,m-1})^{-1} \, \text{Vol}(B_t^{b,m}) \right) dt \ .
\tag{12.3}
$$

Proof. From the general solution (10.4) of the Poisson equation (10.3) (resp. (10.2)) we get

$$
E(r) = \kappa + \int_0^r \left(\text{Vol}(S_t^{b,m-1})^{-1} \left(\tilde{C} - \text{Vol}(B_t^{b,m}) \right) \right) dt \ ,
\tag{12.4}
$$

where κ is an arbitrary constant of integration. In order for the solution to be non-singular (continuous) at $r = 0$ we need $\tilde{C} = 0$, and the boundary condition $E(R) = 0$ gives the then unique solution in (12.3). $\qquad\square$

Remark 12.3. A nice property of the radial symmetry of the functions $E_R^{b,m}(r)$ is displayed by

$$
E_R^{b,m}(r) = E_R^{b,m}(0) - E_r^{b,m}(0) \ ,
\tag{12.5}
$$

which states that the mean exit time from any point in the ball is given by suitable mean exit times for Brownian motions starting from the center of the ball.

We are going to present at this point a maximum principle for subharmonic functions which sustains our comparison results for the mean exit time function. Given a domain (connected open set) Ω in M, a function $u \in C^2(\Omega)$ is *harmonic* (resp. *subharmonic*) if $\Delta u = 0$ (resp. $\Delta u \geq 0$) on Ω. Then

Theorem 12.4 ([J]). *Let Ω be a smooth domain of a Riemannian manifold M. Consider a subharmonic function $u \in C^2(\Omega) \cap C(\overline{\Omega})$. Then the maximum principle says:*

(i) *if u achieves its maximum in Ω then u is constant,*

(ii) *if there is $p_0 \in \partial\Omega$ such that $u(p) < u(p_0)$ for any $p \in \Omega$ then*

$$
\frac{\partial u}{\partial \nu}(p_0) = \langle \nabla u, \partial \nu \rangle (p_0) > 0 \ ,
$$

where ν denotes the outer unit normal along $\partial\Omega$.

From the maximum principle it is clear that any subharmonic function on a compact manifold M without boundary must be constant.

When we consider a complete immersed submanifold P^m in an ambient Riemannian manifold N^n, our previous comparison machinery may now be applied to give comparison results for the mean exit time functions on minimal extrinsic balls D_R.

Theorem 12.5 ([Ma6], [MaP1]). *We let $D_R(p)$ denote a minimal extrinsic ball of P^m in N and assume that $K_N \leq b \leq 0$. Then the mean exit time $E_R(x)$ from the point x in D_R satisfies the inequality*

$$E_R(x) \leq E_R^{b,m}(r(x)) \text{ for all } x \in D_R , \qquad (12.6)$$

where, as usual, we have denoted $r(x) = \operatorname{dist}_N(p, x)$.

Suppose we assume instead that $K_N \geq b \geq 0$. Then the mean exit time $E_R(x)$ from the point x in D_R satisfies the reverse inequality

$$E_R(x) \geq E_R^{b,m}(r(x)) \text{ for all } x \in D_R . \qquad (12.7)$$

If equality is attained in one of the inequalities (12.6) or (12.7) for just one point x in the interior of D_R then equality holds for all points in D_R. If furthermore $b \neq 0$, then the standard rigidity conclusions hold.

Remark 12.6. Again this theorem represents evidence that the Laplacian operates 'swiftly' on minimal submanifolds in negatively curved ambient spaces and that it operates 'slowly' in positively curved ambient spaces.

Proof. In view of the fact that we are forced to assume $\tilde{C} = 0$ for obtaining the solution $E_R^{b,m}(r)$ to the Poisson equation in the comparison ball $B_R^{b,m}$, we obtain from Corollary 10.2 (equation (10.6)) the same inequality as the one used in the proof of Theorem 11.3, case $b \leq 0$, i.e.,

$$\Delta^P E_R^{b,m}(r(x)) \leq -1 = \Delta^P E_R(x) . \qquad (12.8)$$

But then

$$\Delta^P \left(E_R^{b,m}(r(x)) - E_R(x) \right) \leq 0 , \qquad (12.9)$$

so that the difference $E_R^{b,m}(r(x)) - E_R(x)$ is a superharmonic function on D_R. Since this function furthermore vanishes on the boundary ∂D_R, it follows, applying Theorem 12.4, that the difference function is nonnegative in all of D_R. This proves the first statement of the theorem. If the equality occurs for one point in the interior of D_R, then the mean value inequality for superharmonic functions on minimal extrinsic balls gives equality in all of D_R (see, e.g., [CheLY2], Corollary 1 p. 1052 and the generalization in [Ma3], Corollary A p. 481), and then the standard rigidity conclusions follow as above. □

Exercise 12.7. Show the second part, i.e., (12.7) of the theorem, e.g., by using one of the other three Laplace comparison settings alluded to in Section 5.

Remark 12.8. It should be noted that we do not get (by these methods, at least) any comparison of exit times from minimal extrinsic annuli, i.e., from domains of type $A_{\rho,R}(p) = D_R(p) - D_\rho(p)$ – not even in constant curvature ambient spaces – unless, of course, the annulus is totally geodesic.

Exercise 12.9. What is the mean exit time from a point in the totally geodesic annulus $\Lambda_{\rho,R}^{b,m}(\tilde{p}) = B_R^{b,m}(\tilde{p}) - B_\rho^{b,m}(\tilde{p})$ in $\mathbb{K}^m(b)$?

13. The Poisson Hierarchy

Let us consider a smooth precompact domain Ω in a Riemannian manifold M. The mean exit time function $E(x)$ defined above is but only the first in a sequence of functions $\{u_1 = E, u_2, \dots\}$ defined inductively on Ω as follows:

$$\Delta u_1 + 1 = 0 \text{ on } \Omega$$
$$u_1|_{\partial\Omega} = 0 , \tag{13.1}$$

and for $k \geq 2$,

$$\Delta u_k + k\, u_{k-1} = 0 \text{ on } \Omega$$
$$u_k|_{\partial\Omega} = 0 . \tag{13.2}$$

We will refer to this sequence of solutions as *the Poisson hierarchy* for Ω – as it is named in [DLD].

The Poisson hierarchy in turn gives rise to the so-called p-moment spectrum of the domain Ω for every $p > 0$ as follows – see [H, KD, KDM, McDa, Dy]:

Definition 13.1. We consider a smooth precompact domain Ω in a Riemannian manifold M. For each p the L^p-moment spectrum of Ω is defined as the sequence of integrals:

$$\mathcal{A}_{p,\,k}(\Omega) = \left(\int_\Omega (u_k(x))^p \, dV \right)^{1/p} , \quad k = 1,\dots,\infty . \tag{13.3}$$

Following [McDa] we will write shorthand as follows for the L^1-spectrum:

$$\mathcal{A}_k(\Omega) = \mathcal{A}_{1,\,k}(\Omega) . \tag{13.4}$$

The name of this *moment*-spectrum follows from the fact, that the k'th element in the spectrum is precisely the k'th moment of the mean exit time function, see [CLD]:

$$\mathcal{A}_k(\Omega) = \int_\Omega \mathbb{E}[\tau^k] \, dV . \tag{13.5}$$

Not surprisingly, the invariants $\mathcal{A}_k(\Omega)$ are related to the Dirichlet heat kernels $H(p, x, t)$ in Ω that we encountered in Section 6 – see, e.g., [DLD] and [CLD]. In fact, if we define the *heat content* function $\mathcal{H}$ for Ω as the following double integral over the domain:

$$\mathcal{H}(t) = \int_\Omega \int_\Omega H(z, x, t) \, dV_z \, dV_x , \tag{13.6}$$

then that function of $t > 0$ gives directly the Poisson hierarchy as follows:

$$\mathcal{A}_k(\Omega) = k \cdot \int_0^\infty t^{k-1} \cdot \mathcal{H}(t)\, dt \tag{13.7}$$

and eventually – via the Mellin transform of the heat content – also the original defining equations (13.1) for the Poisson hierarchy functions u_k.

The first element $\mathcal{A}_1(\Omega)$ has historically received special interest, since it is classically well known as the *torsional rigidity of* Ω; this is because when $\Omega \subseteq \mathbb{R}^2$, $\mathcal{A}_1(\Omega)$ represents the torque required for a unit angle of twist per unit length when twisting an elastic beam of uniform cross section Ω, (see [Ba] and [PS]). Clearly, the torsional rigidity $\mathcal{A}_1(\Omega)$ plays a significant role in the mean exit time moment spectrum as does the first positive Dirichlet eigenvalue in the Dirichlet spectrum. Much research already deals with bounds for the first Dirichlet eigenvalue and for the torsional rigidity, respectively, in terms of geometric invariants associated with the underlying domain. See, e.g., [BerGM], [Cha1], [Cha2] concerning the first Dirichlet eigenvalue, and [Ba], [PS], [BBC], and [BG] concerning the torsional rigidity.

The most relevant token of interest in these problems is given by the St. Venant torsion problem. This problem is a precise analog of the Rayleigh conjecture about the *fundamental tone* of a membrane – that amongst all homogeneous membranes of the same area, the circular one has the lowest first Dirichlet eigenvalue. In 1856, Saint-Venant conjectured similarly that among all cross sections with a given area, the circle has maximum torsional rigidity. A first proof of this conjecture was given by G. Pólya in 1948, (see [Po2] and also [PS]), using the method of symmetrizations due to J. Steiner. The proof of the Rayleigh conjecture in Riemannian geometry is based on the Faber–Krahn inequality, which in turn is based on the isoperimetric inequality. This inequality is thus an important connection between the bounds for the first Dirichlet eigenvalue and the geometry of the manifold.

In the paper [McDa], P. McDonald combines both techniques, using now Schwarz symmetrization and the Faber–Krahn inequality, in order to establish relations between the L^1-moment spectrum of a given domain and the geometry of the manifold. He thereby obtains upper bounds for the L^1-moment spectrum of smoothly bounded domains in a Riemannian manifold provided that all the domains satisfy a certain type of isoperimetric condition. Again, these conditions are formulated in comparison with the real space forms of constant curvature $\mathbb{K}^n(b)$, $b \in \mathbb{R}$ (see Theorem 1.2 in [McDa]).

During the last 20 years or so the intricate relationships between the Dirichlet spectrum and the mean exit time moment spectrum for given domains in Riemannian manifolds have been vividly explored, see, e.g., [McDa, McDb, CLD, CGL, BFNT, BBV, BBC] and references therein. Our own results and estimates in this direction have been worked out after the appearance of the first edition of these notes, see [HMPa, HMPb, HMPc]. They have mainly been directed towards establishing inequalities and comparison results concerning elements of the moment spectrum – including specific rigidity results – of the type already illustrated in the previous sections, namely by using transplanted distance functions as a key auxiliary tool. Before we go into a brief description of some of the results alluded to above, we can speculate that further work in this interesting cross-fertilizing field of research may eventually answer a Kac type question regarding the moment spectrum: Although we are not able to pose a question as prolific, intuitively crisp, and suggestive as Mark Kac's original question concerning the Dirichlet spectrum in [Kac], we may nevertheless ask: *Does the mean exit time moment spectrum of a domain Ω in a Riemannian manifold determine the shape of Ω up to isometry?*

We mention briefly some results in this direction: First we introduce a sequence of invariants that play a significant role for the interesting bounds obtained for the Dirichlet eigenvalues in terms of the moment spectrum. See [DLD, CLD] and references therein.

Definition 13.2. Let us consider a smooth precompact domain Ω in a Riemannian manifold M. Let λ denote an eigenvalue in the Dirichlet spectrum $\text{spec}(\Omega)$ of Ω and let E_λ be the eigenspace corresponding to λ. Then we define a_λ^2 to be the square of the L^2-norm of the orthogonal projection of the constant function 1 on the eigenspace E_λ. Moreover, if $a_\lambda^2 > 0$ we will say that λ belongs to the $*$-spectrum of Ω, $\text{spec}^*(\Omega) \subset \text{spec}(\Omega)$.

Upper bounds on $\lambda_n(\Omega)$ are found in [CLD]:

Theorem 13.3 ([CLD], Theorem 1.1). *Let F denote the following function on the pairs of positive integers (ℓ, n):*

$$F(\ell, n) = \frac{\mathcal{A}_\ell(\Omega)}{\ell!} - \left(\sum_{\substack{\nu \in \text{spec}^*(\Omega) \\ \nu < \lambda_n(\Omega)}} a_\nu^2 \cdot \left(\frac{1}{\nu} \right)^\ell \right). \tag{13.8}$$

Then the following estimate holds for all integers k:

$$\lambda_n(\Omega) \leq \frac{F(2k-1, n)}{F(2k, n)}. \tag{13.9}$$

Moreover, if $\lambda_n(\Omega)$ itself belongs to the $*$-spectrum of Ω then

$$\lambda_n(\Omega) = \lim_{k \to \infty} \frac{F(2k-1, n)}{F(2k, n)}. \tag{13.10}$$

Corresponding lower bounds are obtained in [DLD], again in terms of the key auxiliary projections a_λ^2 and the moment spectrum:

Theorem 13.4 ([DLD], Theorem 1.3). *With notation as above, we let λ denote any Dirichlet eigenvalue in $\text{spec}(\Omega)$. Then, for every integer k:*

$$\lambda \geq \left(\frac{k! \cdot a_\lambda^2}{\mathcal{A}_k(\Omega)} \right)^{1/k}. \tag{13.11}$$

Returning now to the main comparison methodology/philosophy of the present notes we indicate but two results from [HMPb, HMPa, HMPc], which are obtained by comparing (again via transplanted distance functions) the moment spectrum of metric balls and of their equivalents in suitable comparison spaces, respectively, see also [MaP6].

As it happens with the first Poisson equation, we also know explicitly the full Poisson hierarchy corresponding to a geodesic ball B_R^b in the space forms $\mathbb{K}^m(b)$ of constant curvature b:

Proposition 13.5 ([HMPb]). *Let $\tilde{u}_k^b$, $k = 1, \ldots, \infty$, be the Poisson hierarchy defined on the geodesic R-ball B_R^b in a space form $\mathbb{K}^m(b)$ of constant curvature b.*

Then the functions $\tilde{u}_k^b$ are radial and given by the expressions

$$\tilde{u}_1^b(r) = \int_r^R \frac{\int_0^t w_b^{m-1}(s)\,ds}{w_b^{m-1}(t)}\,dt, \tag{13.12}$$

and

$$\left(\tilde{u}_k^b\right)'(r) = -k\,\frac{\int_0^r w_b^{m-1}(s)\tilde{u}_{k-1}^b(s)\,ds}{w_b^{m-1}(r)}. \tag{13.13}$$

Therefore,

$$\frac{\mathcal{A}_k(B_R^b)}{\mathrm{Vol}(\partial B_R^b)} = -\frac{1}{k+1}\,\tilde{u}_{k+1}^{b\,'}(R)\,, \tag{13.14}$$

where

$$w_b(r) = \begin{cases} \frac{1}{\sqrt{b}}\sin(\sqrt{b}\,r) & \text{if } b > 0 \\ r & \text{if } b = 0 \\ \frac{1}{\sqrt{-b}}\sinh(\sqrt{-b}\,r) & \text{if } b < 0\,. \end{cases} \tag{13.15}$$

We then have the following:

Theorem 13.6. *Let B_R^N be a p-centered geodesic ball of a complete Riemannian manifold N^n with sectional curvatures bounded from below (respectively from above) by $b \in \mathbb{R}$. For the comparison we let B_R^b denote the geodesic R-ball in $\mathbb{K}^n(b)$. Then the exit moment spectra of B_R^N and B_R^b satisfy the following isoperimetric type inequalities*

$$\frac{\mathcal{A}_k(B_R^N)}{\mathrm{Vol}(\partial B_R^N)} \geq (\leq)\,\frac{\mathcal{A}_k(B_R^b)}{\mathrm{Vol}(\partial B_R^b)}\,. \tag{13.16}$$

Equality in (13.16) for some k implies that B_R^N is isometric to the ball B_R^b and hence again that equality is attained for all k and for every smaller p-centered extrinsic ball in N^n.

Proof. As the functions $\tilde{u}_k^b$ are radial functions, we transplant them to the R-ball B_R^N defining $\tilde{u}_k : B_R^N \to \mathbb{R}$ as $\tilde{u}_k(x) := \tilde{u}_k^b(r(x))$ for all $x \in B_R^N$. We assume first that the sectional curvatures of N satisfy $K_N \geq b$, and consider an orthonormal basis $\{X_1, \ldots, X_n = \nabla^N r\}$ of the tangent space TN. Then we obtain, using equation (3.3) and Propositions 3.8 and 13.5 that

$$\Delta^N \tilde{u}_k \geq \Delta^{\mathbb{K}^m(b)} \tilde{u}_k^b = -k \tilde{u}_{k-1}^b = -k \tilde{u}_{k-1} . \qquad (13.17)$$

Now we are going to prove *inductively* that if we denote by u_k the solutions of the hierarchy of boundary value problems on B_R^N given by (13.2), then $\tilde{u}_k \leq u_k$ on B_R^N.

For $k = 1$, since $\tilde{u}_0^b$ is assumed to be identically 1, inequality (13.17) gives us that

$$\Delta^N \tilde{u}_1 \geq -1 = \Delta^N u_1 ,$$

so $\Delta^N(\tilde{u}_1 - u_1) \geq 0$ on B_R^N and $\tilde{u}_1 - u_1 = 0$ on ∂B_R^N. Applying the Maximum Principle we conclude that $\tilde{u}_1 \leq u_1$ on B_R^N.

Suppose now that $\tilde{u}_k \leq u_k$ on B_R^N, then as a consequence of inequality (13.17) we get

$$\Delta^N \tilde{u}_{k+1} \geq -(k+1) \tilde{u}_k \geq -(k+1) u_k = \Delta^N u_{k+1} ,$$

and $\tilde{u}_{k+1} - u_{k+1} = 0$ on ∂B_R^N, so applying again the Maximum Principle we have $\tilde{u}_{k+1} \leq u_{k+1}$.

Summarizing, we have so far: $\tilde{u}_k \leq u_k$ and $\Delta^N \tilde{u}_k \geq \Delta^N u_k$ on B_R^N for all $k \geq 1$. Taking these inequalities into account and applying the Divergence theorem, we then get

$$\begin{aligned}
\mathcal{A}_k(B_R^N) &= \int_{B_R^N} u_k \, dV = -\frac{1}{k+1} \int_{B_R^N} \Delta^N u_{k+1} \, dV \\
&\geq -\frac{1}{k+1} \int_{B_R^N} \Delta^N \tilde{u}_{k+1} \, dV \\
&= -\frac{1}{k+1} \int_{\partial B_R^N} \langle \nabla^N \tilde{u}_{k+1}, \nabla^N r \rangle \, dA \\
&= -\frac{1}{k+1} \tilde{u}_{k+1}^{b'}(R) \, \mathrm{Vol}(\partial B_R^N) .
\end{aligned}$$

Then, we apply Proposition 13.5 again to get the result. For $K_N \leq b$ the proof follows along the same lines upon reversing the relevant inequalities correspondingly. $\square$

Remark 13.7. The main Theorems 1.2 and 1.3 in [HMPb] give similar but more general comparison and rigidity results, namely for extrinsic balls in minimal submanifolds and with warped product model spaces (defined in Section 15 below) as comparison spaces. The isoperimetric flavour of (13.16) is directly comparable with the aforementioned result of P. McDonald in [McDa, Theorem 1.2].

A corresponding bound on the first Dirichlet eigenvalue by the moment spectrum is reported in [HMPc]. Here, for transparency, we restrict the setting to be that of a metric ball in constant curvature. The expression for the first eigenvalue of a geodesic ball in constant curvature is given in terms of the mean exit time moment spectrum for the geodesic ball. Indeed, we get the following:

$$\frac{k\,u_{k-1}(0)}{u_k(0)} \le \lambda_1(B_R^b) \le \frac{k\,\mathcal{A}_{k-1}(B_R^b)}{\mathcal{A}_k(B_R^b)}\;,$$

for all $k \ge 1$ (see (13.2) and (13.3)), where the above bounds improve as k increases. We emphasize that the exact value $\lambda_1(B_R^b)$ is obtained in the limit:

Theorem 13.8. *Let $B_R^b(p)$ be the geodesic ball of radius R centered at the pole p in $\mathbb{K}^n(b)$. Then the first eigenvalue of the ball can be expressed as the following limits of exit time moment data:*

$$\lambda_1(B_R^b) = \lim_{k\to\infty} \frac{k\,u_{k-1}(0)}{u_k(0)} = \lim_{k\to\infty} \frac{k\,\mathcal{A}_{k-1}(B_R^b)}{\mathcal{A}_k(B_R^b)}\;,$$

where u_k are the solutions of the boundary value problem (13.2) defined on the geodesic R-ball B_R^b, and $\mathcal{A}_k(B_R^b)$ is the corresponding k-moment of B_R^b. Moreover, the radial function $g_\infty(r) := \lim_{k\to\infty} \frac{u_k(r)}{u_k(0)}$ is a Dirichlet eigenfunction for the first eigenvalue $\lambda_1(B_R^b)$.

We note also that in [BGJ] the authors generalize this result to bounded domains Ω in the more general setting of weighted manifolds – see Section 21 for the definition and other applications of weighted manifolds.

14. Capacity Comparison

The notions of capacity and effective resistance of, say, a minimal extrinsic annulus $A_{\rho,R} = D_R - D_\rho$ of P^m in N are defined as follows, (see Section 21 for a more general definition of capacity):

Definition 14.1 (See, e.g., [Gri1]). Let $\Psi : A_{\rho,R} \to \mathbb{R}_+ \cup \{0\}$ denote the harmonic function which satisfies

$$\begin{cases} \Delta^P \Psi(x) = 0 \text{ for all } x \in A_{\rho,R} \\ \quad \Psi(x) = 0 \text{ for all } x \in \partial D_\rho \text{ and} \\ \quad \Psi(x) = 1 \text{ for all } x \in \partial D_R\,. \end{cases} \qquad (14.1)$$

Then

$$\mathrm{Cap}^P(A_{\rho,R}) = \int_{\partial D_\rho} |\nabla^P \Psi|\, dA$$
$$= \int_{\partial D_R} |\nabla^P \Psi|\, dA\,, \qquad (14.2)$$

and

$$\mathrm{R_{eff}}^P(A_{\rho,R}) = (\mathrm{Cap}(A_{\rho,R}))^{-1}\,. \qquad (14.3)$$

The capacity $\mathrm{Cap}^P(A_{\rho,R})$ is also called the *Newtonian capacity* of the capacitor (D_ρ, D_R). We refer to Sections 15 and 21 for a more detailed and general exhibition of this concept. The harmonic function equation (14.1) with the given boundary conditions is precisely the Euler–Lagrange equation for the alternative energy expression of the capacity:

$$\mathrm{Cap}^P(A_{\rho,R}) = \inf_{\phi \in \mathbb{L}(A_{\rho,R})} \int_{A_{\rho,R}} |\nabla^P \phi|^2 \, dV , \tag{14.4}$$

where $\mathbb{L}(A_{\rho,R})$ denotes the set of locally Lipschitz functions on P with compact support in $\overline{A}_{\rho,R}$ which satisfy $0 \le \phi \le 1$ and $\phi_{|\partial D_R} = 1$, $\phi_{|\partial D_\rho} = 0$.

The capacitary energy is minimized by the harmonic solution Ψ, whose energy is precisely the flux given in equation (14.2). Indeed, an application of Green's formula (6.4), equation (6.7), gives:

$$\begin{aligned}
\int_{A_{\rho,R}} |\nabla^P \Psi|^2 \, dV &= -\int_{A_{\rho,R}} \Psi \Delta^P \Psi \, dV + \int_{\partial A_{\rho,R}} \Psi |\nabla^P \Psi| \, dA \\
&= 0 - \int_{\partial D_\rho} \Psi |\nabla^P \Psi| \, dA + \int_{\partial D_R} \Psi |\nabla^P \Psi| \, dA \\
&= 0 - 0 + \int_{\partial D_R} |\nabla^P \Psi| \, dA \\
&= \mathrm{Cap}^P(A_{\rho,R}) .
\end{aligned} \tag{14.5}$$

The notion of capacity is an intrinsic concept, but when we consider a submanifold P isometrically immersed in an ambient manifold N, we denote the capacity of the extrinsic capacitors (D_ρ, D_R) as $\mathrm{Cap}^P(A_{\rho,R})$. However, when P^m is a totally geodesic submanifold in $N = \mathbb{K}^n(b)$, and when we consider a purely intrinsic situation, (as in the next Section 15 and in Section 21), we shall use Cap to denote the capacity of the corresponding intrinsic capacitors.

When P^m is a totally geodesic submanifold in $N = \mathbb{K}^n(b)$, we have the totally geodesic annulus $\Lambda_{\rho,R}^{b,m} = B_R^{b,m} - B_\rho^{b,m}$. Its capacity is given by the following

Proposition 14.2. *The capacity of the totally geodesic annulus $\Lambda_{\rho,R}^{b,m}$ of constant curvature b is given by*

$$\mathrm{Cap}(\Lambda_{\rho,R}^{b,m}) = \left(\int_\rho^R \mathrm{Vol}(S_r^{b,m-1})^{-1} \, dr \right)^{-1} . \tag{14.6}$$

Proof. For $q = 0$ the general solution to the Laplace equation on the level of $\Gamma(r)$, i.e., equation (10.2), is given by

$$\Gamma(r) = \tilde{C} \cdot \mathrm{Vol}(S_r^{b,m-1})^{-1} , \tag{14.7}$$

so that the solution to the Dirichlet problem (14.1) for the annulus $\Lambda_{\rho,R}^{b,m}$ is determined by

$$\Psi_{\rho,R}^{b,m}(r) = \tilde{C} \cdot \int_\rho^r \mathrm{Vol}(S_t^{b,m-1})^{-1} \, dt , \tag{14.8}$$

where now $\tilde{C}$ has to be determined so that the outer boundary condition $\Psi_{\rho,R}^{b,m}(R) = 1$ is also satisfied. This is clearly accomplished by

$$\tilde{C} = \left(\int_\rho^R \mathrm{Vol}(S_r^{b,m-1})^{-1} \, dr \right)^{-1} . \tag{14.9}$$

The capacity is then obtained as follows

$$\begin{aligned}
\mathrm{Cap}(\Lambda_{\rho,R}^{b,m}) &= \int_{\partial B_\rho^{b,m}} |\nabla^P \Psi_{\rho,R}^{b,m}(r)| \, dA \\
&= \left(\Psi_{\rho,R}^{b,m} \right)'(\rho) \int_{S_\rho^{b,m-1}} |\nabla^P r| \, dA \\
&= \Gamma(\rho) \, \mathrm{Vol}(S_\rho^{b,m-1}) \tag{14.10} \\
&= \tilde{C} \\
&= \left(\int_\rho^R \mathrm{Vol}(S_r^{b,m-1})^{-1} \, dr \right)^{-1} . \qquad \square
\end{aligned}$$

Corollary 14.3. *If $(b < 0$ and $m \geq 2)$ or $(b = 0$ and $m \geq 3)$ then we have for every fixed $\rho > 0$:*

$$\lim_{R \to \infty} \mathrm{Cap}(\Lambda_{\rho,R}^{b,m}) > 0 , \tag{14.11}$$

so that under the given assumptions the effective resistance of the annuli stays finite when the out-radius goes to infinity.

Exercise 14.4. Prove the corollary and calculate the limits in (14.11) explicitly for $b = 0$ and for all $m \in \{2, 3, \dots\}$. The following by now well-known identity may still be useful

$$\mathrm{Vol}(S_r^{b,m-1}) = \mathrm{Vol}(S_1^{0,m-1}) \cdot Q_b(r)^{m-1} . \tag{14.12}$$

Corollary 14.5. *In all cases we have for fixed $R > 0$ that*

$$\lim_{\rho \to 0} \mathrm{Cap}(\Lambda_{\rho,R}^{b,m}) = 0 , \tag{14.13}$$

so that the effective resistance (i.e., the reciprocal of the effective capacity) of the annuli go to infinity when the in-radius goes to 0.

We are now ready to compare the constant curvature annuli with the corresponding annuli of minimal submanifolds.

Theorem 14.6 ([MaP4]). *We let $A_{\rho,R}(p)$ denote a minimal extrinsic annulus of P^m in N (centered at p) and assume that $K_N \leq b$, $b \in \mathbb{R}$. Then the capacity of $A_{\rho,R}(p)$ satisfies the inequality*

$$\mathrm{Cap}^P(A_{\rho,R}) \geq \mathrm{Cap}(\Lambda_{\rho,R}^{b,m}) . \tag{14.14}$$

If equality is attained, then all of D_R is a minimal radial cone in N and the standard rigidity conclusions hold for D_R.

FIGURE 8. An extrinsic minimal annulus of Costa's surface and the flat comparison annulus.

Proof. From the proof of Proposition 14.2 we have $\tilde{C} > 0$ for every finite $0 < \rho < R < \infty$. We are then in position to apply Corollary 10.3 as we did in the proof of Theorem 11.1, but now using $q = 0$. In particular we have

$$\Delta^P \Psi^{b,m}_{\rho,R}(r(x)) \geq 0 = \Delta^P \Psi(x) , \tag{14.15}$$

so that the difference $\Psi^{b,m}_{\rho,R}(r(x)) - \Psi(x)$ is a subharmonic function. Since this function vanishes at the boundary $\partial A_{\rho,R} = \partial D_R \cup \partial D_\rho$, we deduce from the maximum principle in Theorem 12.4 that the difference function is nonpositive in all of $A_{\rho,R}$, i.e.,

$$\Psi(x) \geq \Psi^{b,m}_{\rho,R}(r(x)) \text{ for all } x \in A_{\rho,R} , \tag{14.16}$$

and hence along the in-boundary ∂D_ρ, which happens to be a level-hypersurface for both functions in D_R, we get in particular

$$|\nabla^P \Psi(x)| \geq |\nabla^P \Psi^{b,m}_{\rho,R}(r(x))| \text{ for all } x \in \partial D_\rho . \tag{14.17}$$

Therefore

$$\begin{aligned}
\mathrm{Cap}^P(A_{\rho,R}) &= \int_{\partial D_\rho} |\nabla^P \Psi(x)| \, dA \\
&\geq \int_{\partial D_\rho} |\nabla^P \Psi^{b,m}_{\rho,R}(r(x))| \, dA \\
&= \left(\Psi^{b,m}_{\rho,R}\right)'(\rho) \cdot \int_{\partial D_\rho} |\nabla^P r| \, dA \tag{14.18} \\
&\geq \Gamma(\rho) \cdot \mathrm{Vol}(S^{b,m-1}_\rho) \\
&= \tilde{C} \\
&= \mathrm{Cap}(\Lambda^{b,m}_{\rho,R}) ,
\end{aligned}$$

FIGURE 9. Another view of the extrinsic minimal annulus of Costa's surface and the flat comparison annulus.

where we have used the result of Corollary 11.4 on the inner disc D_ρ of (or better 'out of') the annulus:

$$\mathrm{Vol}(S_\rho^{b,m-1}) \leq \int_{\partial D_\rho} |\nabla^P r| \, dA \leq \mathrm{Vol}(\partial D_\rho) . \tag{14.19}$$

If we have equality in all of (14.18), then we get from the equality conclusions in Corollary 11.4 that the extension of the two functions $\Psi(x)$ and $\Psi_{\rho,R}^{b,m}(r(x))$ into $D_\rho - \{p\}$ agree in this inner disc, so that the difference $\Psi_{\rho,R}^{b,m}(r(x)) - \Psi(x)$ is still subharmonic and nonpositive (in fact 0) there. The difference then vanishes identically in all of $D_R - \{p\}$, because otherwise there would be an interior maximum point (of value 0), which contradicts the mean value inequality for subharmonic functions on minimal submanifolds, cf. again, e.g., [CheLY2], Corollary 1 p. 1052 and the generalization in [Ma3], Corollary A p. 481. Therefore we obtain equality in the comparison statement (14.15), so that the annulus $A_{\rho,R}$ and hence all of D_R is radially generated.   □

Corollary 14.7 ([MaP4]). *Suppose that the assumptions of Theorem 14.6 are satisfied. If furthermore N is simply connected and ($b < 0$ and $m \geq 2$) or ($b = 0$ and $m \geq 3$) – so that in particular N is a Cartan–Hadamard manifold – then we have for every fixed $\rho > 0$:*

$$\lim_{R \to \infty} \mathrm{Cap}^P(A_{\rho,R}) > 0 . \tag{14.20}$$

Under the given assumptions the effective resistances of the minimal extrinsic annuli are therefore bounded when the out-radius goes to infinity.

Proof. This follows immediately from Theorem 14.6 together with Corollary 14.3.
□

FIGURE 10. A third view of the extrinsic minimal annulus of Costa's surface and the flat comparison annulus.

15. The Kelvin–Nevanlinna–Royden Criteria for Transience

At this point we list (without proof) an important series of criteria which are equivalent to the property of having 'finite resistance to infinity'. For a thorough discussion and proofs of the equivalences, see, e.g., [LS], [Gri1], and also [Ly.T] and [So]. These latter two references cover the interesting case of locally finite metric graphs, which will also be discussed briefly – by example – in the last section of these notes.

Let (M, g) be a given Riemannian manifold and let $\Omega \subseteq M$ denote a precompact open domain in M. For the more general definition of $\mathrm{Cap}(M - \Omega) = \mathrm{Cap}(\Omega, M)$ and $\mathrm{R}_{\mathrm{eff}}(M - \Omega) = \frac{1}{\mathrm{Cap}(\Omega, M)}$ in the following theorem we refer to Section 21, where this notion is discussed in the general setting of weighted manifolds M.

Theorem 15.1 (After T. Lyons and D. Sullivan, [LS]). *Let (M, g) be a given Riemannian manifold. Then the following conditions are equivalent.*

- *M has* finite resistance *to infinity: There exists in M a precompact open domain Ω, such that*

$$\mathrm{R}_{\mathrm{eff}}(M - \Omega) < \infty \ . \tag{15.1}$$

- *M has* positive capacity*: There exists in M a precompact open domain Ω, such that*

$$\mathrm{Cap}(M - \Omega) > 0 \ . \tag{15.2}$$

- *M is* hyperbolic *in the sense that M admits a Green function*

$$G(x, y) = \int_0^\infty H(x, y, t)\, dt \ < \ \infty \ \text{for all } x \neq y \ , \tag{15.3}$$

where $H(x, y, t)$ is the heat kernel on M.
(Otherwise M is called parabolic.)

- *M is* non-parabolic *in the sense that there exists a non-constant bounded above subharmonic function globally defined on M.*

- *M is* transient *in the sense that there is a precompact open domain Ω, such that the Brownian motion X_t starting from Ω does not return to Ω with*

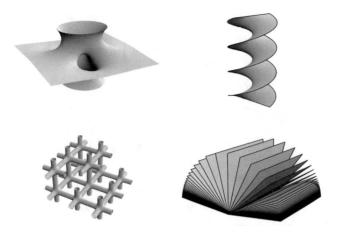

FIGURE 11. A Variety of surfaces in $\mathbb{R}^3$. Two of them have finite resistance to infinity. These two are hyperbolic and transient in the sense of the Kelvin–Nevanlinna–Royden criteria.

probability 1, *i.e.*:

$$\mathbb{P}^x\{\Omega \mid X_t(\Omega) \in \Omega \text{ for some } t > 0\} < 1 \ . \tag{15.4}$$

(*Otherwise M is called* recurrent.)

- (*The Kelvin–Nevanlinna–Royden criterion*): M admits a finite energy flow *vector field* Ψ *such that*

$$\begin{cases} \int_M |\Psi|^2 \, dV & < \infty \ , \\ \int_M |\operatorname{div}(\Psi)| \, dV < \infty \ , & and \\ \int_M \operatorname{div}(\Psi) \, dV & \neq 0 \ . \end{cases} \tag{15.5}$$

As a partial illustration of the specific Kelvin–Nevanlinna–Royden criterion (15.5) for transience we have:

Theorem 15.2 ([LS]). *If the volume $A(r)$ of the geodesic sphere $\partial B_r(p)$ of radius r centered at some given point p in the manifold satisfies*

$$\int_1^\infty (1/A(r)) \, dr = \infty \ ,$$

then there is no vector field which satisfies the Kelvin–Nevanlinna–Royden criterion. Hence the manifold is parabolic.

Proof. Suppose for contradiction, that a vector field W satisfies the conditions in (15.5). Let ν denote the outward pointing unit normal vector field along $\partial B_r(p)$.

Then the Cauchy–Schwarz theorem and the divergence theorem give:

$$
\begin{aligned}
\int_M |W|^2 \, dV &= \int_0^\infty \left(\int_{\partial B_r(p)} |W|^2 \, dA \right) dr \\
&\geq \int_0^\infty \left(\int_{\partial B_r(p)} \langle W, \nu \rangle^2 dA \right) dr \\
&\geq \int_0^\infty \frac{1}{A(r)} \left(\int_{\partial B_r(p)} \langle W, \nu \rangle dA \right)^2 dr \\
&= \int_0^\infty \frac{1}{A(r)} \left(\int_{B_r(p)} \operatorname{div}(W) \, dV \right)^2 dr = \infty ,
\end{aligned}
\tag{15.6}
$$

and this is the desired contradiction. $\qquad \square$

We may now reformulate Corollary 14.7 as follows:

Corollary 15.3 ([MaP4]). *Suppose we have a standard situation, i.e., P^m is a complete minimally immersed submanifold of a Cartan–Hadamard manifold N and $K_N \leq b$. If ($b < 0$ and $m \geq 2$) or ($b = 0$ and $m \geq 3$) then P^m is hyperbolic, i.e., Brownian motion on P is transient.*

16. Surfaces of Revolution

In the case of general 2D surfaces of revolution with a pole p and polar coordinates (r, θ) centered at p, the metric is $ds^2 = dr^2 + g(r)^2 d\theta^2$ for some nonnegative function $g(r)$ with $g(0) = 0$ and $g'(0) = 1$. The curvature at the point (r, θ) is independent of θ and is given by

$$
K(r) = -\frac{1}{g(r)} \frac{\partial^2 g(r)}{\partial r^2} .
\tag{16.1}
$$

Lemma 16.1 (Milnor, [Mi]). *A given complete surface M of revolution with pole p is conformally diffeomorphic to a disc if and only if*

$$
\int_1^\infty \frac{1}{g(r)} \, dr < \infty .
\tag{16.2}
$$

Proof. Following Milnor we simply observe, that if we map the point (r, θ) from M to the complex number $R(r)e^{i\theta}$ in $\mathbb{R}^2$, where

$$
R(r) = \exp \left(\int_1^r \frac{1}{g(s)} \, ds \right) ,
\tag{16.3}
$$

then the radial stretching factor

$$
\frac{d}{dr} R(r) = R(r)/g(r)
$$

is precisely the same as the 'circumferential' stretching factor

$$2\pi R(r)/2\pi g(r) \;=\; R(r)/g(r)\,,$$

so that the map is conformal. When $r \to 0$ the map is still bounded and conformal, so that the apparent singularity at 0 is removable. In short we have constructed a smooth conformal map which is defined on all of M and whose image in $\mathbb{R}^2$ is either the finite disc of radius $\exp\left(\int_1^r 1/g(s)\,ds\right)$ or the entire plane depending on whether this integral is finite or not. □

Milnor used this precise condition to interpret the influence of curvature on the conformal type of a surface of revolution:

Theorem 16.2 ([Mi]). *We consider two cases:*

(A) *If the curvature function satisfies*

$$K(r) \;\geq\; -1/(r^2 \log r) \;=\; K_0(r) \quad \text{for } r \text{ large},$$

then the surface is conformal to the plane.
The surface is of parabolic *type.*

(B) *If, on the other hand there is an* $\epsilon > 0$ *such that*

$$K(r) \;\leq\; -(1+\epsilon)/(r^2 \log r) \;=\; K_\epsilon(r) \quad \text{for } r \text{ large},$$

then the surface is conformal to a disk.
The surface is of hyperbolic *type.*

Proof. The proof is by way of comparing the function $g(r)$ from the metric of M with a function $g_\epsilon(r)$ which has corresponding curvature function $K_\epsilon(r)$ for $\epsilon \geq 0$ and for sufficiently large r, say $r > 1$.

For the case **(A)** (i.e., $\epsilon = 0$) we may use $g_0(r) = r \log r$ and assume without lack of generality, that $g(1) < g_0(1)$ and that $g'(1) < g_0'(1)$; otherwise we multiply the metric of M by a suitable constant which does not change the conformal type of M. Then it follows from the assumption $K(r) \geq K_0(r)$ (which is an assumption on the relative second derivatives of $g(r)$ and $g_0(r)$), that $g(r) < g_0(r)$ for all $r \geq 1$.

Exercise 16.3. Show this inequality.

We conclude, that the integral $\int_1^\infty (1/g(r))\,dr$ is infinite (because it is already infinite with $g_0(r)$ in place of $g(r)$), so that by Lemma 16.1 M is conformal to the plane.

For the case **(B)** (i.e., $\epsilon > 0$) we may use the comparison function $g_\epsilon(r) = r(\log r)^{1+\epsilon}$. The curvature of the corresponding metric is now $K_\epsilon(r)$ and the corresponding test integral is finite. Using a similar argument as above, but with the reversed inequalities we then get $g(r) > g_\epsilon(r)$ for all $r \geq 1$. But then $\int_1^\infty (1/g(r))\,dr$ is finite (because it is already finite with $g_\epsilon(r)$ in place of $g(r)$), so that again by Lemma 16.1 M is conformal to a disc. □

An important feature of conformal mappings (at least in dimension 2), is encoded into the following general theorem, see, e.g., [GroKM] or [BerGM].

FIGURE 12. A conformal representation of the Helicoid. The map is given by: $\phi(u,v) = (\sinh(u)\cos(v), \sinh(u)\sin(v), v)$.

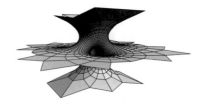

FIGURE 13. A rough conformal representation of Costa's minimal surface.

Theorem 16.4. *Let (P^m, g^P) be a locally conformally flat Riemannian manifold, i.e., the metric g^P is proportional to the flat metric g_0 in $\mathbb{R}^m$ via a positive smooth function ϕ*

$$g^P = \phi \cdot g_0 = \phi \cdot \langle\,,\,\rangle_{g_0}, \tag{16.4}$$

where $\phi > 0$. Then the corresponding Laplacians of a given smooth function ψ are related as follows

$$\phi \Delta^P \psi = \Delta^{R^m}\psi + \left(\frac{m}{2} - 1\right)\langle \nabla\psi, \nabla\log\phi\rangle_{g_0}. \tag{16.5}$$

Exercise 16.5. The theorem is clearly stated in a rather sloppy way here. For example there is no explicit reference to the conformal map in question, and the exact domains for the functions are not explicitly stated. The exercise is to fix it.

In dimension 2 every conformal map therefore preserves harmonic functions, and harmonic functions are precisely the potentials that we use for calculating the capacity of annular domains.

Corresponding domains have thus the same capacity and effective resistance:

Proposition 16.6. *The capacity* $\mathrm{Cap}(\Omega)$ *and thus the effective resistance* $\mathrm{R}_{\mathrm{eff}}(\Omega)$ *of an annular domain* Ω *of a two-dimensional Riemannian manifold is an invariant under conformal diffeomorphisms.*

Proof. Exercise. $\square$

Example 16.7 (The helicoid). We illustrate the conformally related annular domains in the Helicoid and in the plane, respectively, by Figure 14.

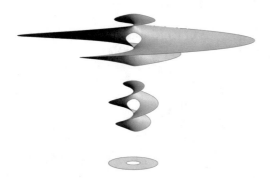

FIGURE 14. At the floor: A flat annulus $\Lambda_{\rho,R}^{0,2}$. In the middle: A minimal extrinsic comparison annulus $A_{\rho,R}$ cut out from a Helicoid P in $N = \mathbb{R}^3$. At the top: Another larger part of the same Helicoid, namely the image of $\Lambda_{\rho,R}^{0,2}$ by the conformal mapping $\phi : \Lambda_{\rho,R}^{0,2} \to \mathbb{R}^3$ determined by $\phi(u,v) = (\sinh(u)\cos(v), \sinh(u)\sin(v), v)$. The heuristically reasonable moral of this constellation is the following: $\mathrm{R}_{\mathrm{eff}}(\phi(\Lambda_{\rho,R}^{0,2})) = \mathrm{R}_{\mathrm{eff}}(\Lambda_{\rho,R}^{0,2}) > \mathrm{R}_{\mathrm{eff}}(A_{\rho,R})$, or equivalently: $\mathrm{Cap}(\phi(\Lambda_{\rho,R}^{0,2})) = \mathrm{Cap}(\Lambda_{\rho,R}^{0,2}) < \mathrm{Cap}(A_{\rho,R})$.

17. Warped Products

Warped products are generalized manifolds of revolution. We refer to [GreW] and [O'N] for excellent treatments of warped product manifolds. Let (B^q, g_B) and (F^p, g_F) denote two Riemannian manifolds and let $f : B \to \mathbb{R}_+$ be a positive real function on B. Then the product manifold $M = B \times F$ is considered. We denote the respective projections onto the factors by $\pi : M \to B$ and $\sigma : M \to F$ respectively. The metric g on M is then defined by the following f-modified (warped) product metric

$$g = \pi^*(g_B) + (f \circ \pi)^2 \sigma^*(g_F). \tag{17.1}$$

Definition 17.1. The Riemannian manifold $(M, g) = (B \times F, g)$ is called a warped product with warping function f, base manifold B and fiber F.

The Laplacian of a function on a warped product may be expressed in terms of the local restrictions of the function to the base and fiber. Here we only consider the simplified case of a real function Ψ on the warped product M, which is the lift of a real function ψ on the base, i.e., we assume that $\Psi = \psi \circ \pi$.

Then the M-gradient of Ψ is the lift to M of the B-gradient of ψ, and thus we get the following formula for the Laplacian of Ψ:

Proposition 17.2.

$$\Delta^M \Psi = \Delta^B \psi + \frac{\dim(F)}{f} \cdot \langle \nabla^B \psi, \nabla^B f \rangle_B . \qquad (17.2)$$

Proof. Following the notation of [O'N] we consider a frame field in M: $\{E_i\}_{i=1}^{i=m} = \{X_i\}_{i=1}^{i=\dim(B)} \cup \{V_i\}_{i=1}^{i=\dim(F)}$. Let $Z = \nabla^M \Psi$. Then

$$\Delta^M \Psi = \operatorname{div}^M Z$$

$$= \sum_{i=1}^{m} \langle D_{E_i}^M Z, E_i \rangle$$

$$= \sum_{i=1}^{\dim(B)} \langle D_{X_i}^M Z, X_i \rangle + \sum_{i=1}^{\dim(F)} \langle D_{V_i}^M Z, V_i \rangle$$

$$= \operatorname{div}^M \nabla^B \psi + \sum_{i=1}^{\dim(F)} \frac{Z(f)}{f} \cdot \langle V_i, V_i \rangle \qquad (17.3)$$

$$= \Delta^B \psi + \sum_{i=1}^{\dim(F)} \frac{1}{f} \cdot Z(f)$$

$$= \Delta^B \psi + \frac{\dim(F)}{f} \cdot Z(f)$$

$$= \Delta^B \psi + \frac{\dim(F)}{f} \cdot \langle \nabla^B \psi, \nabla^B f \rangle_B . \qquad \square$$

Corollary 17.3. *Suppose that B is one-dimensional. Then*

$$\langle \nabla^B \psi, \nabla^B f \rangle_B(t) = \psi'(t) \cdot f'(t) , \qquad (17.4)$$

so that

$$\Delta^M \Psi = \psi''(t) + \dim(F) \cdot \psi'(t) \cdot \frac{f'(t)}{f(t)} . \qquad (17.5)$$

Remark 17.4. This is the warped product generalization of the Laplacian acting on radially symmetric functions in the constant curvature space forms $\mathbb{K}^m(b)$. In particular we note that the functions $h_b(t)$ (and $f_b(t)$, respectively) from that context corresponds precisely to the relative derivative $\frac{f'(t)}{f(t)} = \frac{d}{dt} \ln(f(t))$ in the present context of warped products. This is consistent with the fact that the constant mean curvature of every fiber $\pi^{-1}(t)$ in the warped product is $\frac{d}{dt} \ln(f(t))$. The warped product spaces represent, in short, interesting comparison spaces as

a generalization of the constant curvature comparison spaces that we have so far
been considering.

In particular we note the following

Corollary 17.5. *For a warped product with one-dimensional base B the capacity of
any 'annulus' $\pi^{-1}([a, b])$ only depends on the warping function over the interval
$[a, b]$ and not on the topology or geometry of the fiber F.*

This corollary is, of course, most significant for warped products with fiber-
dimension larger than 1.

18. Answering the Questions in the Appetizer

Proposition 18.1. *The Helicoid is parabolic.*

Proof. The Helicoid is conformally diffeomorphic to the plane, so that it is parabolic according to Proposition 16.6. This has already been alluded to for the finite
annuli in Figure 14 .   □

For the hyperboloid case we use a theorem due to Ichihara, see [I1], which is
a generalization of a previous result due to Blanc and Fiala from 1942, see [BlF].

Theorem 18.2 ([I1]). *If a two-dimensional manifold M has finite total absolute
curvature, then it is parabolic.*

Proof. The curvature of M is (within the cut locus from a given point p) given
in terms of polar coordinates (r, θ) from p by the relative second-order radial
derivative of the spherical part $g(r, \theta)$ of the metric $ds^2 = dr^2 + g^2(r, \theta)d\theta^2$
(compare with the rotationally symmetric case, equation (16.1))

$$K(r, \theta) = -\frac{1}{g(r, \theta)} \frac{\partial^2 g(r, \theta)}{\partial r^2} . \tag{18.1}$$

By assumption we have

$$\int_M |K| dV = \int_{S^1} \left(\int_0^{\mu(\theta)} \left| \frac{\partial^2 g(r, \theta)}{\partial r^2} \right| dr \right) d\theta < \infty , \tag{18.2}$$

where $\mu(\theta)$ denotes the distance from p to the cut locus of p in the direction θ.
We let $C(\theta)$ denote the radial part of the integrand, which by assumption is then
also finite:

$$C(\theta) = \int_0^{\mu(\theta)} \left| \frac{\partial^2 g(r, \theta)}{\partial r^2} \right| dr < \infty . \tag{18.3}$$

Within the cut locus (i.e., for all $r < \mu(\theta)$) we have

$$\left| \frac{\partial g(r, \theta)}{\partial r} \right| \leq \int_0^r \left| \frac{\partial^2 g(s, \theta)}{\partial s^2} \right| ds + \left| \frac{\partial g(0, \theta)}{\partial r} \right|$$
$$= \left| \frac{\partial^2 g(s, \theta)}{\partial s^2} \right| ds + 1 . \tag{18.4}$$

Since $g(0, \theta) = 0$, we therefore have the inequalities

$$g(r, \theta) \leq \int_0^r \left| \frac{\partial g(s, \theta)}{\partial s} \right| ds \leq (C(\theta) + 1)r . \tag{18.5}$$

The fundamental idea is now to choose a sequence of 'radial' gradient fields $\nabla \Psi_R$ on the surface and show that the 'capacitary' energy of the fields goes to 0 when pushing them through larger and larger annular domains, $A_{1,R}$ (as $R \to \infty$) of the surface. This implies the parabolicity of the surface.

We define Ψ_R as follows:

$$\Psi_R(x) = \psi_R(\text{dist}_M(p, x)) , \tag{18.6}$$

where

$$\psi_R(r) = \log\left(\tfrac{R}{r}\right) / \log(R) \text{ for } r \in]1, R[. \tag{18.7}$$

Then in the annulus $A_{1,R}$ consisting of the points in the distance range $[1, R]$ from p we get

$$\Psi_R \in \mathbb{L}(A_{1,R}) , \tag{18.8}$$

and we thus have the following estimate for the energy of $\nabla \Psi_R$:

$$\begin{aligned}
\text{Cap}(A_{1,R}) &\leq \int_{B_R(p)} \langle \nabla \Psi_R, \nabla \Psi_R \rangle dV \\
&= \int_{S^1} \left(\int_1^{\min(R, \mu(\theta))} \left(\frac{d\psi_R(r)}{dr} \right)^2 g(r, \theta)dr \right) d\theta \\
&= \int_{S^1} \left(\int_1^{\min(R, \mu(\theta))} \frac{1}{r^2} g(r, \theta)dr \right) / (\ln(R))^2 d\theta \qquad (18.9) \\
&\leq \int_{S^1} \left(\int_1^R \frac{1}{r} (1 + C(\theta))dr \right) / (\ln(R))^2 d\theta \\
&\leq \left(2\pi + \int_M |K| dV \right) / (\ln(R)) .
\end{aligned}$$

The last expression clearly goes to 0 when R goes to infinity. Hence the surface is parabolic. □

Remark 18.3. Infinite total absolute curvature does not imply hyperbolicity. Just consider the surface of revolution around the x-axis with generator curve, say, $y = 1 + \epsilon \sin(x)$. The helicoid is another example.

Remark 18.4. Infinite topology does not in itself guarantee hyperbolicity either. An example which shows this is the half-infinite 'Jacobs ladder', which, when viewed from a far distance looks like a straight half-line, and is therefore parabolic.

We are now ready to show that the hyperboloids in $\mathbb{R}^3$ are parabolic and that the hyperbolic paraboloids are parabolic as well. Note the unfortunate dichotomy of terms in these statements. To avoid any further confusion we recall the definitions of these surfaces:

Definition 18.5. A hyperboloid in $\mathbb{R}^3$ is defined as a level surface of the function $f(x, y, z) = ax^2 + by^2 + cz^2$, where a, b, and c are non-zero without all having the same sign, and a hyperbolic paraboloid is defined as the graph-surface in $\mathbb{R}^3$ of the function $g(x, y) = ax^2 + by^2$, where $ab < 0$.

Proposition 18.6. *The hyperboloids and the hyperbolic paraboloids in $\mathbb{R}^3$ all have finite total absolute curvature and are therefore parabolic according to Ichihara's theorem.*

Exercise 18.7. Show this claim about finite total absolute curvature of the surfaces considered in Proposition 18.6. You may like to use that the curvature of the graph surface $z = h(x, y)$ is given by the formula

$$K(x, y) = \left(h_{xx}h_{yy} - h_{xy}^2 \right) / \left(1 + h_x^2 + h_y^2 \right)^2 \tag{18.10}$$

– or use the formula (16.1) for the curvature of a rotationally symmetric surface, respectively.

Among the classical characterizations of parabolic manifolds we must also mention a key result due to L. Ahlfors, which has been extended by several authors, see [Ahl, Gri1]: In a rotationally symmetric manifold (warped product manifold) with r-spheres S_r the condition $\int^\infty \frac{dr}{Vol(S_r)} = \infty$ is equivalent to parabolicity. Other examples and extensions of Ahlfors' theorem are obtained by Ichihara, see [I1], for Riemannian manifolds with Ricci curvature bounded from below. Ichihara also proved a non-parabolicity criterion assuming sectional curvatures bounded from above (see [I1, I2]).

19. Sufficient Conditions for Parabolicity and Hyperbolicity

Definition 19.1. Let Ω_0 denote a fixed smooth and bounded domain in M. The Ω_0-rooted isoperimetric profile of M is the non-decreasing function denoted by

$$\phi_{M,\Omega_0} : [\mathrm{Vol}(\Omega_0), \mathrm{Vol}(M)[\ \mapsto \ \mathbb{R}_+ \ ,$$

which is defined as follows:

$$\phi_{M,\Omega_0}(t) = \inf\{ \operatorname{Area}(\partial\Omega) \mid$$
$$\Omega \text{ is a smooth relatively compact domain in } M , \tag{19.1}$$
$$\Omega \supset \Omega_0 , \quad \text{and } \mathrm{Vol}(\Omega) \geq t \} .$$

Theorem 19.2 (J.L. Fernandez, [F]). *The manifold (M, g) is hyperbolic if*

$$\int_{\mathrm{Vol}(\Omega_0)}^{\mathrm{Vol}(M)} \phi_{M,\Omega_0}(t)^{-2}dt < \infty . \tag{19.2}$$

Remark 19.3. The volume of M is necessarily infinite, cf. Theorem 19.6.

Proof of Theorem 19.2. We assume for contradiction that $\mathrm{Cap}(M - \Omega_0) = 0$. In effect this has the following consequence: We let $\{\Omega_0 \subseteq \Omega_1 \subseteq \Omega_2 \subseteq \cdots\}$ denote a compact exhaustion of M by smooth domains based at Ω_0 such that $\overline{\Omega}_n \subseteq \Omega_{n+1}$ and $\bigcup_{n=0}^{\infty} \Omega_n = M$. Further for each n we let Ω_n denote the capacity minimizing solution to the Dirichlet problem

$$\begin{cases} \Delta\Omega_n = 0 \ \text{ in } \Omega_n - \overline{\Omega}_0 \\ \quad \Omega_n = 0 \ \text{ on } \partial\Omega_0 \\ \quad \Omega_n = 1 \ \text{ on } \partial\Omega_n \ . \end{cases} \tag{19.3}$$

Then by assumption we have that $\Omega_n \to 0$ uniformly on compact subsets of $M - \Omega_0$ and that $\langle \nabla\Omega, \nu \rangle \to 0$ as well, where ν denotes the outer normal vector field on $\partial\Omega_0$ – because otherwise we could construct a harmonic capacity potential function on M with nonvanishing energy from the sequence of solutions Ω_n.

Suppose ϵ and δ are given real numbers such that $0 \leq \epsilon < \delta < 1$. Then for each n we consider that subdomain $\Omega_{n,\epsilon,\delta}$ of $\Omega_n - \overline{\Omega}_0$ where Ω_n takes values in between ϵ and δ, i.e., $\Omega_{n,\epsilon,\delta} = \{x \in \Omega_n - \overline{\Omega}_0 \mid \epsilon < \Omega_n(x) < \delta\}$.

Since Ω_n is harmonic in $\Omega_{n,\epsilon,\delta}$, we get from Green's theorem (6.4):

$$\int_{\Omega_{n,\epsilon,\delta}} |\nabla\Omega_n|^2 \, dV = \alpha_n(\delta - \epsilon) \ , \tag{19.4}$$

where

$$\alpha_n = \int_{\partial\Omega_0} \langle \nabla\Omega_n, \nu \rangle \, dA \ . \tag{19.5}$$

We observe that under the assumptions above we then have that $\alpha_n \to 0$ for $n \to \infty$. On the other hand from the co-area formula, Theorem 6.1, we also have in this setting

$$\int_{\Omega_{n,\epsilon,\delta}} |\nabla\Omega_n| \, dV = \int_{\epsilon}^{\delta} A(t) \, dt \ , \tag{19.6}$$

where $A(t)$ denotes the volume of the implicitly given surface in $\Omega_{n,\epsilon,\delta}$ where Ω_n attains the value t.

The Schwarz inequality now gives

$$\left(\int_{\epsilon}^{\delta} A(t)dt\right)^2 \leq (\delta - \epsilon) \cdot \alpha_n \cdot \mathrm{Vol}(\Omega_{n,\epsilon,\delta}) \ , \tag{19.7}$$

so that

$$\left(\frac{1}{\delta - \epsilon} \int_{\epsilon}^{\delta} A(t)dt\right)^2 \leq \alpha_n \cdot \frac{\mathrm{Vol}(\Omega_{n,\epsilon,\delta})}{\delta - \epsilon} \ . \tag{19.8}$$

We therefore have for $V_n(\delta) = \mathrm{Vol}(\Omega_{n,0,\delta})$ the following differential inequality:

$$A(\delta)^2 \leq \alpha_n \cdot \frac{d}{d\delta} V_n(\delta) \ . \tag{19.9}$$

If we use $\phi(t)$ as shorthand for $\phi_{M,\Omega_0}(t)$ we have by definition

$$A(\delta) + A(0) \geq \phi(V_n(\delta)) , \tag{19.10}$$

and hence from the differential inequality

$$(\phi(V_n(\delta)) - A(0))^2 \leq \alpha_n V_n'(\delta) . \tag{19.11}$$

Since $\Omega_n \to 0$ we also have, say, $V_n(\frac{1}{2}) \to \text{Vol}(M)$ for $n \to \infty$.

By assumption we also have $\phi(\text{Vol}(M - \Omega_0)) > 2A(0)$ so that for some sufficiently large $n > n_0$ and for $\delta > \frac{1}{2}$ we get from equation (19.11)

$$\frac{1}{4}\phi(V_n(\delta))^2 \leq \alpha_n V_n'(\delta) , \tag{19.12}$$

so that finally an integration shows

$$\frac{1}{8} \leq \alpha_n \int_{\frac{1}{2}}^1 \frac{V_n'(\delta)d\delta}{\phi(V_n(\delta))^2} \leq \alpha_n \int_{\text{Vol}(\Omega_0)}^{\text{Vol}(M)} \phi(t)^{-2}dt , \tag{19.13}$$

which contradicts the assumption that $\alpha_n \to 0$ for $n \to \infty$.   □

Remark 19.4. A similar theorem may be formulated and proved for locally finite graphs. See Thomassen's work [T1], where he obtains a very similar result for graphs with quite different techniques, but with the strong extra conclusion that the graph in fact contains a transient tree under the assumption of a square integrable reciprocal isoperimetric profile function for the graph.

Remark 19.5. As already alluded to above no manifold with finite volume can satisfy the condition in Fernandez' theorem. In fact a finite volume manifold is parabolic by the following theorems due to Grigoryan, Karp, Lyons and Sullivan, and Varopoulos; see [Gri1] Section 7.2 for a proof of these results using the capacity condition (15.2) for transience from Theorem 15.1.

Theorem 19.6. *Let $V(q,r)$ denote the volume of the geodesic ball centered at q in M and with radius r. If there exists a point q such that*

$$\int_1^\infty \frac{r}{V(q,r)}dr = \infty , \tag{19.14}$$

then M is parabolic.

Similarly and motivated again by the study of warped product manifolds, the weak growth of the boundary of the geodesic balls is also sufficient to guarantee parabolicity as observed already in Theorem 15.2, which, of course, may be reformulated as follows:

Theorem 19.7. *Let $S(q,\rho)$ denote the volume of the boundary of the geodesic ball centered at q in M and with radius ρ. If there exists a point q such that*

$$\int_1^\infty \frac{1}{S(q,r)}dr = \infty , \tag{19.15}$$

then M is parabolic.

Remark 19.8. Neither (19.14) nor (19.15) is a necessary condition for parabolicity. This is shown by nice examples in [Gri1]. However, as already alluded at the end of Section 18, (19.14) and (19.15) characterizes parabolicity when we consider the space forms of constant curvature $\mathbb{K}^m(b)$, or, more generally, rotationally symmetric spaces, (see Section 23).

20. Hyperbolicity of Spacelike Hypersurfaces

In this section we shall study the hyperbolicity of spacelike hypersurfaces with controlled mean curvatures in spacetimes with timelike sectional curvatures bounded from below. For that purpose, and motivated by the Riemannian results above, we naturally apply the analysis of the Lorentzian distance function, which was already presented in Section 4.

First of all, we recall (see Theorem 15.1), that a Riemannian manifold M is hyperbolic (non-parabolic) if and only if there exists a non-constant subharmonic function which is bounded from above and globally defined on M.

Remark 20.1.

   i) This definition is equivalent to the fact that there exists a non-constant positive superharmonic function globally defined on M. To see the equivalence, observe that if f is a non-constant subharmonic function bounded from above on M, then choosing $C > \max_M f$ we obtain $C - f$ a non-constant positive superharmonic function. Conversely, if f is a non-constant positive superharmonic function on M, then $-f$ is a non-constant subharmonic function bounded from above on M.

   ii) On the other hand, the existence of a non-constant positive superharmonic function f globally defined on M is equivalent to the existence of a non-constant bounded (from above and from below) subharmonic function globally defined on M.

   To see the equivalence observe that if f is a non-constant bounded (from above and from below) subharmonic function on M, then choosing $C > \max_M f$ we obtain $C - f$ a non-constant positive superharmonic function. Conversely, if f is a non-constant positive superharmonic function on M, then $0 < \frac{f}{\sqrt{1+f^2}} \leq 1$ determines a non-constant bounded (from above and from below) subharmonic function.

As a consequence of our previous results, (see Section 4), we have the following

Theorem 20.2 ([AHP]). *Let N^{n+1} be an $(n+1)$-dimensional spacetime, $n \geq 2$, such that $K_N(\Pi) \geq b$ for all timelike planes in N. Assume that there exists a point $p \in N^{n+1}$ such that $\mathcal{I}^+(p) \neq \emptyset$, and let $\psi : \Sigma \to N^{n+1}$ be a spacelike hypersurface with $\psi(\Sigma) \subset \mathcal{I}^+(p)$. Let us denote by u the function d_p along the hypersurface, and assume that $u \leq \pi/(2\sqrt{-b})$ if $b < 0$. Then*

(i) *If the future mean curvature of Σ satisfies*

$$H \leq \frac{2\sqrt{n-1}}{n} f_b(u) \quad (\text{with } H < f_b(u) \text{ at some point of } \Sigma \text{ if } n = 2) \qquad (20.1)$$

then Σ is hyperbolic.

(ii) *If $b = 0$ and $H \leq 0$, then Σ is hyperbolic.*

(iii) *If $b > 0$ and $H \leq \frac{2\sqrt{n-1}}{n}\sqrt{b}$, then Σ is hyperbolic.*

In particular, every maximal hypersurface contained in $\mathcal{I}^+(p)$ (and satisfying $u < (\pi/2\sqrt{-b})$ if $b < 0$) is hyperbolic.

Proof. In order to prove (i), observe that u is a non-constant positive function defined on Σ. Otherwise, Σ would be an open piece of the level set given by $d_p = u$, with $\Delta^\Sigma u = 0$ and $\nabla^\Sigma u = 0$, and by Proposition 4.8 its mean curvature would be $H \geq f_b(u)$, which cannot happen because of (20.1). Now we apply Proposition 4.8 to get

$$\Delta^\Sigma u \leq -f_b(u)(n + |\nabla^\Sigma u|^2) + nH\sqrt{1 + |\nabla^\Sigma u|^2} .$$

Let us consider now the function

$$\phi(x) = \frac{n + x^2}{n\sqrt{1 + x^2}}, \quad \text{with} \quad x \geq 0 .$$

Observe that $x = \sqrt{n-2}$ is a minimum of this function with $\phi(\sqrt{n-2}) = 2\sqrt{n-1}/n$.

Therefore

$$\frac{2\sqrt{n-1}}{n} \leq \frac{n + |\nabla^\Sigma u|^2}{n\sqrt{1 + |\nabla^\Sigma u|^2}} .$$

Since $f_b(u) \geq 0$ (recall that we assume $u \leq \pi/(2\sqrt{-b})$ if $b < 0$), then our hypothesis on H implies that

$$H \leq \frac{2\sqrt{n-1}}{n} f_b(u) \leq \frac{f_b(u)(n + |\nabla^\Sigma u|^2)}{n\sqrt{1 + |\nabla^\Sigma u|^2}} .$$

That is,

$$nH\sqrt{1 + |\nabla^\Sigma u|^2} \leq f_b(u)(n + |\nabla^\Sigma u|^2)$$

which yields $\Delta^\Sigma u \leq 0$. As a consequence, u is a non-constant positive superharmonic function on Σ and hence it is hyperbolic.

To prove (ii) and (iii), simply observe that $f_0(u) = 1/u > 0$ and $f_b(u) = \sqrt{b}\coth(\sqrt{b}u) > \sqrt{b}$ on Σ. $\qquad\square$

Remark 20.3. This result could be considered as the Lorentzian counterpart of Theorem 7.6 (see also Corollary 15.3). Moreover, it also has a version in so-called weighted manifolds, which is introduced below, see Corollary 26.2.

21. Weighted Riemannian Manifolds

The notion of *weighted manifolds* generalizes the notion of Riemannian manifolds, so we will use this section for two purposes: Firstly to describe the natural questions that arise from considering the type problem, i.e., parabolicity versus hyperbolicity, in this new and wider context; secondly to give an account of some of the new results, (which includes the Riemannian cases) in this field from the last 20 years.

We shall present some of the main results obtained in the preprints [HPR1] and [HPR2] by C. Rosales and the first and third named authors, with due reference to the previous works and results concerning the weighted setting of several authors, [Ba, Mo, Gri2, Gri3, GriMa, W, Q, WW, PRRS], as well as to the work of the second and the third named authors in the last years, [MaP4, MaP5, MaP7, EP], which are concerned with the Riemannian case and also continues the results presented in the preceding sections.

Let (N, g) be a complete Riemannian manifold. A *density* e^h, where $h : N \to \mathbb{R}$ a smooth function on N, is used to put a controlled weight on the Hausdorff measures associated to the Riemannian metric. In particular, for any Borel set $E \subseteq N$, and any C^1 hypersurface $P \subseteq N$, the *weighted volume* of E and the *weighted area* of P are given by

$$\mathrm{Vol}_h(E) := \int_E dV_h = \int_E e^h \, dV, \quad \mathrm{Vol}_h(P) := \int_P dA_h = \int_P e^h \, dA, \qquad (21.1)$$

where dV and dA denote the Riemannian elements of volume and area, respectively, (see [Mo, Ch. 18] for an introduction to this generalization of Riemannian geometry).

The density function determines not only generalizations of volume and area, but also generalizations of some key differential operators on Riemannian manifolds. We will denote the background metric of the Riemannian manifold by $g = \langle ., . \rangle$. We define the *weighted Laplacian* or *h-Laplacian* of a function $u \in C^2(N)$ as in [Gri2, Sect. 2.1],

$$\Delta_h u := \Delta u + \langle \nabla h, \nabla u \rangle, \qquad (21.2)$$

where Δ and ∇ stand for the Laplace–Beltrami operator and the gradient of a function, respectively.

Given a domain (connected open set) Ω in N, a function $u \in C^2(\Omega)$ is *h-harmonic* (resp. *h-subharmonic*) if $\Delta_h u = 0$ (resp. $\Delta_h u \geq 0$) on Ω. As in the unweighted setting there is a strong maximum principle and a Hopf boundary point lemma for *h*-subharmonic functions, that reads exactly as Theorem 12.4, but replacing subharmonic functions by *h*-subharmonic functions.

Also, as in the unweighted context, from this weighted maximum principle it is clear that any *h*-subharmonic function on a compact manifold N must be constant, and it is natural to wonder what happens in the non-compact case. This question leads to the notion of weighted parabolicity, (see Theorem 15.1 for the Riemannian approach).

Definition 21.1. A weighted manifold N is *weighted parabolic* or *h-parabolic* if any h-subharmonic function which is bounded from above must be constant. Otherwise we say that N is *weighted hyperbolic* or *h-hyperbolic*.

This notion of weighted parabolicity generalizes the concept of Riemannian parabolicity, so it is natural to look for generalizations of the results which relate this invariant property with the geometry of the underlying weighted manifold.

22. Weighted Capacities

The notion of weighted capacity $\mathrm{Cap}_h(K)$ of a precompact set K plays a key role in the study of the type problem concerning the parabolicity-versus-hyperbolicity question. In particular, the h-parabolicity of N is again characterized by the fact that $\mathrm{Cap}_h(K) = 0$ for any/some compact set $K \subseteq M$ with non-empty interior (see [GriSa]). In this section we present the notion of weighted capacity, thereby revising and extending to the weighted context the notions presented in Section 14.

Next, we will recall how the h-parabolicity of manifolds can be characterized by means of weighted capacities. Let $\Omega \subseteq N$ be an open set and $K \subseteq \Omega$ a compact set. The *weighted Newtonian capacity* of the *capacitor* (K, Ω) is defined by

$$\mathrm{Cap}_h(K, \Omega) = \inf_{\phi \in \mathbb{L}(K, \Omega)} \int_\Omega |\nabla \phi|^2 \, dV_h , \tag{22.1}$$

where $\mathbb{L}(K, \Omega)$ denotes the set of locally Lipschitz functions on M with compact support in $\Omega - \overline{K}$ which satisfy $0 \leq \phi \leq 1$ and $\phi_{|\partial K} = 0$, $\phi_{|\partial \Omega} = 1$. Note that the *extrinsic annulus* $A_{\rho, R} \subseteq P \subseteq N$ in Section 9 corresponds to the capacitor (D_ρ, D_R), while the capacitor (B_ρ^N, B_R^N) is an *intrinsic annulus*. When $\Omega = N$, $\mathrm{Cap}_h(K, N) = \mathrm{Cap}_h(K)$ is the h-capacity of K at infinity.

Moreover, if $\Omega \subseteq N$ is precompact, it can be proved, that the *h-capacity* of the compact set K in Ω is given as the following integral, see [Gri1], [Gri2], and equation (14.5) in Section 14:

$$\mathrm{Cap}_h(K, \Omega) = \int_\Omega |\nabla \phi|^2 e^h dV = \int_{\partial K} |\nabla \phi| e^h dA = \int_{\partial K} \frac{\partial u}{\partial \nu} dA_h , \tag{22.2}$$

where the vector field ν is the outer unit normal along $\partial(\Omega \setminus K)$, i.e., the unit normal along ∂K pointing into K, and ϕ is the solution of the Laplace equation on $\Omega - K$ with Dirichlet boundary values:

$$\begin{cases} \Delta_h u = 0 , \\ u \mid_{\partial K} = 0 , \\ u \mid_{\partial \Omega} = 1 . \end{cases} \tag{22.3}$$

In other words, the infimum in (22.1) is attained by the solution to the h-Laplace equation with Dirichlet condition on the boundary (22.3). The function ϕ is called the *h-capacity potential* of the capacitor (K, Ω).

The relation among h-capacity and the h-parabolic or h-hyperbolic character (the h-type) of the manifold is given by the following result, see [Gri3] and Theorem 15.1 in Section 15:

Theorem 22.1. *Let (N, g, e^h) be a weighted manifold. Then, N is h-parabolic (resp. h-hyperbolic) iff N has zero h-capacity, i.e., there exists a compact $D \subseteq N$ with non-empty interior such that $\mathrm{Cap}_h(D, N) = 0$, (resp. N has positive capacity, i.e., there exists a compact $D \subseteq N$ with non-empty interior such that $\mathrm{Cap}_h(D, N) > 0$).*

On the other hand, given $K \subseteq N$ a (pre)compact subset of N, if we consider $\{\Omega_i\}_{i=1}^\infty$ an exhaustion of N by nested and precompact sets, such that $K \subseteq \Omega_i$ for some i, then the h-capacity of K in N, i.e., the h-*capacity at infinity* $\mathrm{Cap}_h(K, N) = \mathrm{Cap}_h(K))$ is given by the following limit:

$$\mathrm{Cap}_h(K, N) = \lim_{i \to \infty} \mathrm{Cap}_h(K, \Omega_i) . \tag{22.4}$$

This definition is independent of the chosen exhaustion.

In view of Theorem 22.1, and the last observation, in order to determine the h-parabolicity of a weighted manifold it suffices to find bounds for the h-capacity of a family of capacitors $\{(K, \Omega_i)\}_{i=1}^\infty$. This is the fundamental idea to obtain a geometric description for the weighted parabolicity in terms of the bounds on the "weighted" curvatures which appear in this theory as generalizations of the standard curvatures defined in the Riemannian setting – see Section 24 below.

23. Weighted Rotationally Symmetric Spaces and the Ahlfors Criterion for Weighted Parabolicity

As already alluded to in Section 17, the works [GreW, Gri1, O'N], provide a complete description of the rotationally symmetric model spaces M_w^m, that we will use to establish our comparison theorems. We remark that these spaces constitute a huge set of comparison spaces that we will now apply in the same way as we used the constant curvature spaces $\mathbb{K}^n(b)$ in Section 3. The spaces $\mathbb{K}^n(b)$ are but particular examples of rotationally symmetric model spaces. We shall need the formal definition of warped products as follows:

Definition 23.1 (see [GreW, Ch. 2], [Gri1, Sect. 3], [Pe, Ch. 3]). A w-model space is a smooth warped product $(M_w^m, g_w) := B^1 \times_w F^{m-1}$ with base $B^1 := [0, \Lambda[\subset \mathbb{R}$ (where $0 < \Lambda \leq \infty$), fiber $F^{m-1} := \mathbb{S}_1^{m-1}$ (the unit $(m-1)$-sphere with standard metric), and warping function $w : [0, \Lambda[\to [0, \infty[$ such that $w(r) > 0$ for all $r > 0$, whereas $w(0) = 0$, $w'(0) = 1$, and $w^{(k)}(0) = 0$ for all even derivation orders k. The point $o_w := \pi^{-1}(0)$, where π denotes the projection onto B^1, is called the *center point* of the model space. If $\Lambda = \infty$, then o_w is a pole of the manifold (recall that a *pole* of a complete Riemannian manifold M is a point $o \in M$ such that the exponential map $\exp_o : T_o M \to M$ is a diffeomorphism).

Example 23.2. The simply connected space forms $\mathbb{K}^m(b)$ of constant sectional curvature b can be constructed as w-models with any given point as center point using the warping functions

$$
w_b(r) := \begin{cases} \frac{1}{\sqrt{b}}\sin(\sqrt{b}\,r) & \text{if } b > 0 \\ r & \text{if } b = 0 \\ \frac{1}{\sqrt{-b}}\sinh(\sqrt{-b}\,r) & \text{if } b < 0 \,. \end{cases}
$$

Note that, for $b > 0$, the function $w_b(r)$ admits a smooth extension to $r = \pi/\sqrt{b}$. For $b \le 0$ any center point is a pole.

In a model space the sectional curvatures for 2-planes containing the radial direction from the center point are determined by the radial function

$$
-\frac{w''(r)}{w(r)} \,.
$$

where $r = \mathrm{dist}_{M_w}(o_w, \cdot)$ is the distance from the center $o_w \in M_w$. Moreover, the mean curvature of the metric sphere of radius r from the center o_w is

$$
\frac{w'(r)}{w(r)} = \frac{d}{dr}\ln(w(r)) \,.
$$

Now we introduce a radial weight $e^{f(r)}$ in M_w^m. A *weighted (w, f)-model space* is a triple $(M_w^m, g_w, e^{f(r)})$ where $e^{f(r)}$ is a radial weight in the w-model (M_w^m, g_w).

We shall denote by B_R^w the geodesic ball of radius R and center the pole o_w. The weighted volumes of geodesic balls and geodesic spheres are then by definition computed as follows:

$$
\mathrm{Vol}_f(B_R^w) = V_0 \int_0^R w^{m-1}(t)\, e^{f(t)}\, dt \,, \tag{23.1}
$$

$$
\mathrm{Vol}_f(\partial B_R^w) = V_0\, w^{m-1}(R)\, e^{f(R)} \,, \tag{23.2}
$$

where V_0 is the volume of the unit sphere S_1^{m-1}.

We will denote by $q_{w,f}(r)$ the corresponding weighted isoperimetric quotient function:

$$
q_{w,f}(r) = \frac{\int_0^r w^{m-1}(t)\, e^{f(t)}\, dt}{w^{m-1}(r)\, e^{f(r)}} \,.
$$

Now we compute the weighted capacity of the capacitor $(\overline{B}_\rho^w, B_R^w)$. Recall that the weighted Dirichlet problem in the annulus $A_{\rho,R}^w := B_R^w \setminus \overline{B}_\rho^w$ is the following:

$$
\begin{cases} \Delta_f^{M_w} u = 0 & \text{on } A_{\rho,R}^w \,, \\ u = 0 & \text{on } \partial B_\rho^w \,, \\ u = 1 & \text{on } \partial B_R^w \,. \end{cases} \tag{23.3}
$$

Theorem 23.3. *Let $A_{\rho,R}^w$ be the annulus of radius ρ and R in a weighted (w, f)-model space. The solution to the Dirichlet problem (23.3) is given by the radial* *function*

$$\Psi_{\rho,R,f}(r) = \left(\int_\rho^r w^{1-m}(s)\, e^{-f(s)}\, ds \right) \left(\int_\rho^R w^{1-m}(s)\, e^{-f(s)}\, ds \right)^{-1}. \tag{23.4}$$

Therefore,

$$\mathrm{Cap}_f(A_{\rho,R}^w) := \mathrm{Cap}_f(\overline{B}_\rho^w, B_R^w) = |\Psi'_{\rho,R,f}(\rho)|\, \mathrm{Vol}_f(\partial B_\rho^w)$$

$$= V_0 \left(\int_\rho^R w^{1-m}(s)\, e^{-f(s)}\, ds \right)^{-1}. \tag{23.5}$$

Proof. The radial function $\Psi(r) = \Psi_{\rho,R,f}(r)$ defined on $A_{\rho,R}^w$ satisfies the first equation in (23.3) if and only if

$$\Psi''(r) + \Psi'(r) \left((m-1)\frac{w'(r)}{w(r)} + f'(r) \right) = 0. \tag{23.6}$$

$\square$

Remark 23.4. Note that by equation (23.2), the f-capacity of the annulus $A_{\rho,R}^w$ can be written in terms of the weighted area of the geodesic spheres as follows:

$$\mathrm{Cap}_f(A_{\rho,R}^w) = \left(\int_\rho^R \frac{dt}{\mathrm{Vol}_f(\partial B_t^w)} \right)^{-1}.$$

As a direct consequence, we have the weighted version of Ahlfors criterion for weighted models, see [Gri2], Section 18, and Section 19.

Corollary 23.5. *Let* $(M_w^m, g, e^{f(r)})$ *be a weighted* (w, f)-*model space. Then* M_w *is* f-*parabolic if and only if* $\mathrm{Cap}_f(B_\rho^w) = 0$ *if and only if*

$$\int_\rho^\infty \frac{ds}{\mathrm{Vol}_f(\partial B_s^w)} = \infty. \tag{23.7}$$

24. Weighted Curvatures

Let us consider a complete non-compact weighted manifold (N^n, g, e^h). Among the several generalizations of the Ricci curvature tensor (which we use to control the Laplacian), the most extensively used are the Bakry–Emery Ricci tensors:

- The *finite* or q-Bakry–Emery Ricci tensor, $q > 0$:

$$\mathrm{Ric}_h^q = \mathrm{Ric} - \mathrm{Hess}\, h - \frac{1}{q}\, dh \otimes dh.$$

- The *infinite* or ∞-Bakry–Emery Ricci tensor:

$$\mathrm{Ric}_h = \mathrm{Ric} - \mathrm{Hess}\, h.$$

Observe that

$$\mathrm{Ric}_h = \mathrm{Ric}_h^q + \frac{1}{q}\, dh \otimes dh \, ,$$

so lower bounds on Ric_h^q implies lower bounds on Ric_h. Here Ric denotes the Ricci tensor in (N, g).

The weighted radial sectional curvatures, (which agree, up to a constant, with the weighted sectional curvatures introduced in [W]), are defined as follows:

Definition 24.1. Consider the radial plane $\sigma_p \subseteq T_pN$ spanned by linearly independent unit vectors X and Y, i.e., $\sigma_p = \mathrm{span}\{X, Y\}$. We define the ∞-*weighted sectional curvature* of σ_p in a weighted Riemannian manifold (N^n, g, e^h) as

$$\mathrm{Sec}_h^\infty(\sigma_p) = K_N(\sigma_p) - \frac{1}{n-1}(\mathrm{Hess}\, h)_p(Y, Y) \, .$$

The *q-weighted sectional curvature* of the radial plane $\sigma_p = \mathrm{span}\{X, Y\}$ is defined by:

$$\mathrm{Sec}_h^q(\sigma_p) = K_N(\sigma_p) - \frac{1}{n-1}(\mathrm{Hess}\, h)_p(X, Y) - \frac{1}{(n-1)q}\, (dh \otimes dh)\, (Y, Y) \, .$$

In the paper [HPR2] an analysis of the intrinsic distance function to a pole in the weighted manifold M^m is carried out. Analogously, as in Sections 3 and 8, the estimates of the Hessian and the weighted Laplacian of the distance from the pole lead to capacity comparison results and parabolicity criteria for weighted manifolds under lower bounds on the Ricci curvatures Ric_h and Ric_h^q.

We also found in [HPR2] bounds for the weighted isoperimetric quotients and volumes of metric balls, along the lines of Section 8, but now from an intrinsic point of view. We must remark that the works [Q, L, Mo2, WW, PRRS] contain previous comparison results that involve weighted volumes and quotients of weighted volumes (although not weighted isoperimetric quotients) when Ric_∞^h is bounded from below.

25. Analysis of Restricted Distance Functions in Weighted Submanifolds

In [HPR1] and [HPR2] the restricted distance to a submanifold immersed in an ambient weighted manifold, is analyzed in the same vein as in Section 3. Let P^m be an m-dimensional submanifold with $\partial P = \emptyset$ properly immersed in a weighted manifold (N^n, g, e^h) with a pole $o \in N$. We consider in P the induced Riemannian metric. We use the notation $\nabla^P u$ and $\Delta^P u$ for the gradient and Laplacian in P of a function $u \in C^2(P)$.

The restriction to P of the weight e^h in N produces a structure of weighted manifold in P. From (21.2) the associated h-*Laplacian* Δ_h^P has the expression

$$\Delta_h^P u = \Delta^P u + \langle \nabla^P h, \nabla^P u \rangle \, ,$$

for any $u \in C^2(P)$. We say that the submanifold P is *h-parabolic* when P is weighted parabolic considered as a weighted manifold. Otherwise we say that P is *h-hyperbolic*. By Theorem 22.1 the h-parabolicity of P is equivalent to $\text{Cap}_h^P(D) = 0$ for some precompact open set $D \subseteq P$, where Cap_h^P denotes the *h-capacity relative to P*. Clearly a compact submanifold P is h-parabolic.

Now we present another necessary ingredient to establish these results: the weighted mean curvature of submanifolds. In the case of two-sided hypersurfaces this was first introduced by Gromov [G], see also [Ba, Ch. 3].

Definition 25.1. The *weighted mean curvature vector* or *h-mean curvature vector* of P is the vector field normal to P given by

$$\overline{H}_h^P := m\overline{H}^P - (\nabla h)^\perp,$$

where $(\nabla h)^\perp$ is the normal projection of ∇h with respect to P and $\overline{H}^P$ is the mean curvature vector of P. This is defined as $m\overline{H}^P := -\sum_{i=1}^{n-m}(\text{div}^P N_i) N_i$, where div^P stands for the divergence relative to P and $\{N_1, \ldots, N_{n-m}\}$ is any local orthonormal basis of vector fields normal to P.

We say that P has *constant h-mean curvature* if $|\overline{H}_h^P|$ is constant on P. If $\overline{H}_h^P = 0$, then P is called *h-minimal*. More generally, P has *bounded h-mean curvature* if $|\overline{H}_h^P| \leq c$ on P for some constant $c > 0$.

For later use we note that the equality

$$\langle m\overline{H}^P, \nabla r \rangle + \langle \nabla^P h, \nabla^P r \rangle = \langle \overline{H}_h^P, \nabla r \rangle + \langle \nabla h, \nabla r \rangle \tag{25.1}$$

holds on $P - \{o\}$. This easily follows from the definition of $\overline{H}_h^P$ and the fact that $\nabla h - (\nabla h)^\perp = \nabla^P h$.

As in Proposition 3.13, some inequalities for the weighted Laplacian of submanifolds have been established under bounds (which are not necessarily constants) on the radial sectional curvatures of the ambient manifold. In the particular case of a weighted (w, f)-model space it can be checked that all the estimates in the next statement become equalities.

Theorem 25.2. *Let (N^n, g, e^h) be a weighted manifold, P^m a submanifold immersed in N, $r: N \to [0, \infty[$ the distance function from a pole $o \in N$, and $w(s)$ a smooth function such that $w(0) = 0$ and $w(s) > 0$ for all $s > 0$.*

If, for any $p \in N - \{o\}$ and any plane $\sigma_p \subseteq T_pN$ containing $(\nabla r)_p$, we have

$$K_N(\sigma_p) \geq (\leq) - \frac{w''(r)}{w(r)},$$

then, for every smooth function $F : (a, b) \to \mathbb{R}$ with $(a, b) \subseteq (0, \infty)$ and $F' \geq 0$, we obtain the inequality

$$\Delta_h^P(F \circ r) \leq (\geq) \left(F''(r) - F'(r) \frac{w'(r)}{w(r)} \right) |\nabla^P r|^2$$

$$+ F'(r) \left(m \frac{w'(r)}{w(r)} + \langle \overline{H}_h^P, \nabla r \rangle + \langle \nabla h, \nabla r \rangle \right)$$

in the points of P where $a < r < b$.

Proof. From the results in, e.g., [MaP7, Pa1], the bound for the radial sectional curvatures of the ambient manifold implies that the Laplacian Δ^P of the modified distance function $F \circ r$ satisfies the inequality

$$\Delta^P(F \circ r) \leq (\geq) \left(F''(r) - F'(r) \frac{w'(r)}{w(r)} \right) |\nabla^P r|^2 + m F'(r) \left(\frac{w'(r)}{w(r)} + \langle \overline{H}^P, \nabla r \rangle \right) .$$

Thus, by the definition of weighted Laplacian, we get

$$\Delta_h^P(F \circ r) \leq (\geq) \left(F''(r) - F'(r) \frac{w'(r)}{w(r)} \right) |\nabla^P r|^2$$

$$+ m F'(r) \left(\frac{w'(r)}{w(r)} + \langle \overline{H}^P, \nabla r \rangle \right) + F'(r) \langle \nabla^P h, \nabla^P r \rangle ,$$

so that the claim follows by using (25.1). □

We must remark here that there are also comparison results for the weighted Laplacian of submanifolds assuming a lower bound on some q-weighted sectional curvatures, (see [HPR2]).

26. Extrinsic Criteria for Weighted Parabolicity

In this section we analyze the weighted parabolicity of a submanifold P immersed in a weighted manifold M as it was done, (in the Riemannian case), in Sections 14, 18 and 19. In this case we consider the restriction to P of the distance function to the pole, (the *extrinsic* distance), and the family of capacitors in the submanifolds, constituted by the *extrinsic balls*.

Our first result is an extension to the weighted setting of previous theorems for Riemannian manifolds in [EP] by Esteve and the third author, and in [MaP4, MaP7]. We note that the particular situations of weighted (w, f)-model spaces were established in Theorems 3.2 and 3.3 of [HPR1].

In [HPR2] a parabolicity criterion is established assuming lower bounds on the q-weighted sectional curvatures of the ambient manifold and upper bounds on the derivatives of the weight and the radial weighted mean curvatures of the submanifold.

Theorem 26.1. *Let (N^n, g, e^h) be a weighted manifold, P^m a non-compact subman-ifold properly immersed in N, $r : N \to [0, \infty[$ the distance function from a pole $o \in N$, and $w(s)$ a smooth function such that $w(0) = 0$, $w'(0) = 1$ and $w(s) > 0$ for all $s > 0$. Suppose that the following conditions are fulfilled:*

(i) *For any $p \in N - \{o\}$ and any plane $\sigma_p \subseteq T_p N$ containing $(\nabla r)_p$, we have*

$$K_N(\sigma_p) \geq (\leq) - \frac{w''(r)}{w(r)} .$$

(ii) *There exists a radius $\rho > 0$ and continuous functions $\psi(s), \varphi(s)$, such that ∂D_ρ is smooth and*

$$\langle \nabla h, \nabla r \rangle \leq (\geq) \psi(r), \quad \langle \overline{H}_h^P, \nabla r \rangle \leq (\geq) \varphi(r) \quad on \ P - D_\rho .$$

(iii) *In $P - D_\rho$ the bounding functions verify*

$$\psi(r) + \varphi(r) \leq (\geq) - m \frac{w'(r)}{w(r)} \quad (balance \ condition) .$$

Then, we obtain

$$\mathrm{Cap}_h^P(D_\rho) \leq (\geq) \frac{\mathrm{Cap}_f(B_\rho^w)}{\mathrm{Vol}_f(\partial B_\rho^w)} \int_{\partial D_R} |\nabla^P r| \, dA_h ,$$

where $\mathrm{Cap}_f(B_\rho^w)$ denotes the weighted capacity of the metric ball B_ρ^w in a weighted (w, f)-model space $(N_w^m, g_w, e^{f(r)})$ with $f(r) := \int_\rho^r (\psi(s) + \varphi(s)) \, ds$ for any $r \geq \rho$. Moreover, if

$$\int_\rho^\infty w^{1-m}(s) \, e^{-f(s)} \, ds = (<) \infty , \tag{26.1}$$

then P is h-parabolic (h-hyperbolic).

Proof. By using Sard's Theorem we can suppose that $\nabla^P r \neq 0$ along ∂D_ρ. Take any number $R > \rho$ such that ∂D_R is smooth. Let us consider the extrinsic annulus $A_{\rho,R}^P := D_R - \overline{D}_\rho$ and the function $\Psi_{\rho,R,f} : [\rho, R] \to \mathbb{R}$ defined in (23.4), i.e., the f-capacity potential of $(\overline{B}_\rho^w, B_R^w)$ in the m-dimensional weighted (w, f)-model space $(N_w^m, g_w, e^{f(r)})$. This function is the solution to the problem (23.3); in particular, it satisfies (23.6). The composition $v := \Psi_{\rho,R,f} \circ r$ defines a smooth function in $A_{\rho,R}^P$.

Since $\Psi'_{\rho,R,f}(r) \geq 0$, by using Theorem 25.2 and the boundedness assumptions (ii) in the statement, we get this comparison in $A_{\rho,R}^P$

$$\Delta_h^P v \leq (\geq) \left(\Psi''_{\rho,R,f}(r) - \Psi'_{\rho,R,f}(r) \frac{w'(r)}{w(r)} \right) |\nabla^P r|^2$$
$$+ \Psi'_{\rho,R,f}(r) \left(\frac{m \, w'(r)}{w(r)} + \varphi(r) + \psi(r) \right) . \tag{26.2}$$

On the other hand, by taking (23.6) and the balance condition in (iii) into account, it follows that

$$\Psi''_{\rho,R,f}(r) - \Psi'_{\rho,R,f}(r) \frac{w'(r)}{w(r)} = -\Psi'_{\rho,R,f}(r) \left(m \frac{w'(r)}{w(r)} + \psi(r) + \varphi(r) \right) \leq (\geq) 0$$

in $A_{\rho,R}^P$. As $|\nabla^P r| \leq 1$, we conclude that

$$\Delta_h^P v \leq (\geq) \, \Psi''_{\rho,R,f}(r) + \Psi'_{\rho,R,f}(r) \left((m-1)\frac{w'(r)}{w(r)} + \varphi(r) + \psi(r) \right) = 0 = \Delta_h^P u \,,$$

where u is the h-capacity potential of the capacitor $(\overline{D}_\rho, D_R)$ in P. Since $u = v$ on $\partial A_{\rho,R}^P$, by applying the weighted version of the maximum principle and the Hopf boundary point lemma in Theorem 12.4, we deduce that $\frac{\partial u}{\partial \nu} > (<) \frac{\partial v}{\partial \nu}$ on ∂D_ρ, where ν is the outer unit normal along $\partial A_{\rho,R}^P$, which coincides with the unit normal $\frac{-\nabla^P u}{|\nabla^P u|} = \frac{-\nabla^P v}{|\nabla^P v|}$ along ∂D_ρ pointing into D_ρ. From (22.2), we obtain

$$\mathrm{Cap}_h^P(D_\rho, D_R) = \int_{\partial D_\rho} |\nabla^P u| \, dA_h \leq (\geq) \int_{\partial D_\rho} |\nabla^P v| \, dA_h$$

$$= |\Psi'_{\rho,R,f}(\rho)| \int_{\partial D_\rho} |\nabla^P r| \, dA_h$$

$$= \frac{\mathrm{Cap}_f(B_\rho^w, B_R^w)}{\mathrm{Vol}_f(\partial B_\rho^w)} \int_{\partial D_\rho} |\nabla^P r| \, dA_h \,.$$

Hence, the desired comparison comes from the above identities by taking limits when $R \to \infty$. Moreover, if (26.1) holds, then $\mathrm{Cap}_f(B_\rho^w) = \mathrm{Cap}_h^P(D_\rho) = 0$ (resp. $\mathrm{Cap}_f(B_\rho^w) > 0$ and $\mathrm{Cap}_h^P(D_R) > 0$), so that P is h-parabolic (resp. h-hyperbolic) by Theorem 22.1. $\qquad\square$

Also, as a consequence of Theorem 26.1 we can extend Theorem 7.6 (see also Corollary 15.3) to a weighted setting.

Corollary 26.2. *Let (N^n, g) be a Cartan–Hadamard manifold, i.e., a complete and simply connected Riemannian manifold such that*

$$K_N(\sigma_p) \leq b \leq 0 \,,$$

for any plane $\sigma_p \subseteq T_pM$ and any point $p \in M$. Denote by $r : N \to [0, \infty[$ the distance function from a fixed point $o \in N$. Given a weight e^h in N, suppose that there exist $m \in \mathbb{N}$ with $m \geq 2$, and constants $\rho, \epsilon > 0$, such that

$$\begin{cases} \langle \nabla h, \nabla r \rangle \geq -\frac{m-2-\epsilon}{r} & \text{in } N - B_\rho \text{ if } b = 0 \,, \\ \langle \nabla h, \nabla r \rangle \geq -(m-1-\epsilon)\sqrt{-b}\coth(\sqrt{-b}\,r) & \text{in } N - B_\rho \text{ if } b < 0 \,. \end{cases}$$

Then, any non-compact h-minimal submanifold P^m properly immersed in N is h-hyperbolic.

Proof. For a submanifold P in the conditions of the statement we check that the hypotheses in Theorem 26.1 are satisfied.

In case $b = 0$ we consider the functions $w, \psi, \varphi : [0, \infty[\to \mathbb{R}$ defined by $w(s) := s$, $\psi(s) := -\frac{m-2-\epsilon}{s}$ and $\varphi(s) := 0$. Observe that

$$\psi(r) + \varphi(r) + m \frac{w'(r)}{w(r)} = \frac{\epsilon + 2}{r} > 0 \,,$$

so that the balance condition holds. On the other hand, a straightforward computation shows that

$$f(s) := \int_\rho^s \psi(t)\, dt = -(m - 2 - \epsilon) \ln\left(\frac{s}{\rho}\right),$$

and so

$$\int_\rho^\infty w^{1-m}(s)\, e^{-f(s)}\, ds = \frac{1}{\rho^{m-2-\epsilon}} \int_\rho^\infty s^{-(\epsilon+1)}\, ds < \infty,$$

which is the condition in (26.1). From here we conclude that P is h-hyperbolic.

In case $b < 0$ we reason in a similar way with the functions $w, \psi, \varphi : [0, \infty[\to \mathbb{R}$ given by $w(s) := \frac{1}{\sqrt{-b}} \sinh(\sqrt{-b}\, s)$, $\psi(s) := -(m - 1 - \epsilon)\sqrt{-b} \coth(\sqrt{-b}\, s)$ and $\varphi(s) := 0$. □

27. The Grigor'yan–Fernandez Criterion for Weighted Hyperbolicity

Our aim in this subsection is to give a weighted version of the hyperbolicity criterion studied in Section 19 (see also [CHS] for a weighted version with a modified isoperimetric profile).

Let M be a Riemannian manifold with a continuous density $f = e^h$. For any open set $\Omega \subseteq M$, the *weighted isoperimetric profile* of Ω is the function $I_{\Omega,h} : [0, V_h(\Omega)] \to \mathbb{R}$ defined by $I_{\Omega,h}(0) = 0$, and

$$I_{\Omega,h}(v) = \inf \{A_h(\Sigma)\,;\, \Sigma \subseteq \overline{\Omega} \text{ is a compact hypersurface enclosing weighted volume } v\},$$

for any $v \in (0, V_h\Omega)]$.

Remark 27.1. Obviously we get the weighted isoperimetric inequality

$$A_h(\Sigma) \geq I_{\Omega,h}(V_h(E)) \geq I_{M,h}(V_h(E)), \tag{27.1}$$

for any open set $E \subseteq \Omega$ with smooth boundary Σ.

Remark 27.2. On the other hand, if we consider the function $h = 0$, and the sets $\Omega_0 = \{p\}$ and $\Omega = M$ in Definition 19.1, then, given $t \geq 0$, the weighted isoperimetric profile of Ω is the $\Omega_0 = \{p\}$-rooted isoperimetric profile of M, namely, $I_{\Omega,h}(t) = \phi_{M,\{p\}}(t)$.

We consider Ω a precompact domain in M and $\psi : \Omega \to \mathbb{R}$ a smooth function such that $\psi(\Omega) = [a, b]$ with $a < b$. Following the notation of Section 6, applying formula (6.2) with $u = gf = ge^h$, and assuming that the set of critical points of ψ, denoted as Ω_0, has measure zero we deduce,

$$\int_\Omega g\, |\nabla\psi|\, dV_h = \int_a^b \left(\int_{\Sigma_t} g\, dA_{h,t}\right) dt$$

$$\frac{d}{dt} V_h(t) = \int_{\Gamma(t)} |\psi(x)|^{-1}\, dA_{h,t}, \tag{27.2}$$

where $V_h(t) = \mathrm{Vol}_h(\Omega(t))$. So we can state the following weighted version of the co-area formula:

Theorem 27.3.

i) *For every integrable function u on $\overline{\Omega}$:*

$$\int_\Omega u \cdot |\nabla \psi|\, dV_h = \int_a^b \left(\int_{\Gamma(t)} u\, dA_{h,t} \right) dt \, . \tag{27.3}$$

where $dA_{h,t}$ *is the weighted volume element of $\Gamma(t)$.*

ii) *The function $V_h(t) := \mathrm{Vol}_h(\Omega(t))$ is a smooth function on the regular values of ψ, where its derivative is given by (assuming that the set of critical points of ψ has measure zero):*

$$\frac{d}{dt} V_h(t) = \int_{\Gamma(t)} |\nabla \psi|^{-1}\, dA_{h,t} \, . \tag{27.4}$$

Remark 27.4. If $g \geq 0$ on Ω and assuming that the set of critical points of ψ, Ω_0, has null measure, we have, using the first equation in (27.2), that

$$\int_\Omega g\, dV_h = \int_{\Omega_0} g\, dV_h + \int_{\Omega - \Omega_0} (g\, |\nabla \psi|^{-1})\, |\nabla \psi|\, dV_h$$
$$= \int_a^b \left(\int_{\Sigma_t} g\, |\nabla \psi|^{-1}\, dA_{h,t} \right) dt, \tag{27.5}$$

We are now ready to prove a lower bound for the capacity via the isoperimetric profile function.

Theorem 27.5. *Let M be a Riemannian manifold with a continuous density $f = e^h$. Consider a capacitor (K, Ω) in M such that the weighted isoperimetric profile $I_{\Omega,h} : [0, V_h(\Omega)] \to \mathbb{R}$ is nondecreasing. Then, we have the inequality*

$$\mathrm{Cap}_h(K, \Omega) \geq \left(\int_{V_h(K)}^{V_h(\Omega)} \frac{dV}{I_{\Omega,h}(v)^2} \right)^{-1},$$

Proof. We follow the Riemannian proof given in [Gri1, Thm. 8.1]. The main ingredients are the isoperimetric inequality in (27.1) and the co-area formula (27.5). By (22.1) it suffices to obtain the desired lower bound on the energy of any $\phi \in C_0^\infty(\Omega)$ with $0 \leq \phi \leq 1$ and $\phi = 1$ on K.

Take $\phi \in C_0^\infty(\Omega)$ with $0 \leq \phi \leq 1$ and $\phi = 1$ on K. Let $\psi := 1 - \phi$. Note that $\psi \in C^\infty(\Omega)$ with $0 \leq \psi \leq 1$, $\psi = 0$ on K and $\psi = 1$ on $\Omega - \mathrm{supp}(\phi)$. For any $t \in (0,1)$, we define

$$E_t := \{ x \in \Omega \, ; \, \psi(x) < t \}, \quad \Sigma_t := \{ x \in \Omega \, ; \, \psi(x) = t \}, \quad m(t) := V_h(E_t).$$

Note that $\Sigma_t \cap \partial\Omega = \emptyset$ for any $t \in (0,1)$ since $\psi = 1$ near $\partial\Omega$. It is clear that $K \subseteq \overline{E_t} \subseteq \Omega$ and $\partial E_t = \Sigma_t$ for any $t \in (0,1)$. Since $E_t \subseteq E_s$ when $t < s$ we get that $m(t)$ is increasing. Moreover, $m(t)$ is bounded and so it has finite limits when $t \to 0$ and $t \to 1$. By the second equation in (27.2), the function $m(t)$ is differentiable a.e. on $(0,1)$ with

$$m'(t) = \int_{\Sigma_t} |\nabla \psi|^{-1} \, dA_{h,t} \quad \text{a.e. on } (0,1). \tag{27.6}$$

Now we prove the claim. By applying (27.5) with $g = |\nabla \psi|^2$ we obtain

$$\int_\Omega |\nabla \phi|^2 \, dV_h = \int_\Omega |\nabla \psi|^2 \, dV_h = \int_0^1 \left(\int_{\Sigma_t} |\nabla \psi| \, dA_{h,t} \right) dt$$

$$\geq \int_0^1 \frac{A_{h,t}(\Sigma_t)^2}{\left(\int_{\Sigma_t} |\nabla \psi|^{-1} \, dA_{h,t} \right)} \, dt \geq \int_0^1 \frac{I_{\Omega,h}(m(t))^2}{m'(t)} \, dt$$

$$\geq \left(\int_0^1 \frac{m'(t)}{I_{\Omega,h}(m(t))^2} \right)^{-1} dt \geq \left(\int_{m(0)}^{m(1)} \frac{dV}{I_{\Omega,h}(v)^2} \right)^{-1} \tag{27.7}$$

$$\geq \left(\int_{V_h(K)}^{V_h(\Omega)} \frac{dV}{I_{\Omega,h}(v)^2} \right)^{-1}.$$

To get the first inequality we have used Hölder's inequality. More precisely

$$A_{h,t}(\Sigma_t) = \int_{\Sigma_t} |\nabla \psi|^{1/2} |\nabla \psi|^{-1/2} \, dA_{h,t}$$

$$\leq \left(\int_{\Sigma_t} |\nabla \psi| \, dA_{h,t} \right)^{1/2} \left(\int_{\Sigma_t} |\nabla \psi|^{-1} \, dA_{h,t} \right)^{1/2}.$$

The second inequality follows from (27.1) and (27.6). The third inequality is a particular case of $\int_0^1 \eta(t) dt \geq (\int_0^1 \eta(t)^{-1} dt)^{-1}$, which holds, by applying Hölder's inequality as above, for any positive function η such that $\eta^{-1} \in L^1(0,1)$. The fourth inequality comes from the change of variables formula since $I_{\Omega,h}$ is nondecreasing. Finally, the last inequality follows since $V_h(K) \leq m(0)$ and $m(1) \leq V_h(\Omega)$.

This proves the claim. $\square$

28. Graphs and Flows

In the two final sections of these notes we now briefly indicate how many of the concepts from the previous sections can be carried over almost verbatim to locally finite graphs and thus give a fruitful alternative understanding of what is going on. We first recall that the dimension of P^m has so far been assumed to be $m > 1$, partly because the intrinsic geometry of geodesic segments is completely trivial. This viewpoint is, of course, altered considerably if we allow geodesic segments to be joined in such a way as to form a *metric geodesic graph* in the ambient space. The analysis of restricted distance functions on *minimal* geodesic graphs in Riemannian manifolds has been studied in [Ma7] following essentially the same lines of reasoning – and the same set of goals – as in the first edition of the present notes.

In view of the new results for (sub-)manifolds that we have reported above, there are then several interesting and pertinent challenges and questions that are calling for similar results to hold true on (minimal) metric graphs in (weighted) manifolds with suitable bounds on their curvatures. Indeed, the notions of Dirichlet spectrum, mean exit time moment spectrum, the weighted capacities, the type problem, etc. are all well defined on such graph-structures. Therefore the quest for finding "good" relations between the Dirichlet spectrum and the moment spectrum in that setting is quite natural. This also holds for the quest of answering the Kac question for either one of the two spectra. These questions can thus be studied following the same lines – or other refined lines with graph theoretic and other discrete tools at our disposal – as in the previous smooth geometric settings. We refer to the papers [McDMa, McDMb, CKD], which contain interesting results in this direction.

Here, however, we will only consider the purely combinatorial structure of graphs, which already by itself can be considered as carrier of such fundamental notions as potential functions, capacity, isoperimetric inequalities, etc., and for which we may also state the type problem, i.e., whether random walk on the graph is transient or recurrent. This latter question will be addressed and discussed in detail via a concrete example in the next section.

We let $G = (V, E)$ denote a finite graph with edge set E and vertex set V. If we associate the resistance value of 1 (Ohm) to every edge and consider the current from one vertex a to another vertex b in G, then we are studying potential theory on finite networks. We refer to [DoyS], [So], and [Wo]. Following [DoyS] we show explicitly the relation between the capacity energy and the harmonic potential in the case of a finite graph.

Definition 28.1. A *flow j* in a graph G from vertex a to vertex b is a function j_{xy} on the space of pairs $(x, y) \in V \times V$ such that

$$\begin{cases} j_{xy} = -j_{yx} \\ \sum_y j_{xy} = 0 \text{ for all } x \neq a, b \\ j_{xy} = 0 \text{ if } x \text{ and } y \text{ are not connected by an edge in } E . \end{cases} \quad (28.1)$$

We denote by j_x the total flow into the vertex x from the outside:

$$j_x = \sum_y j_{xy} .$$

Then $j_x = 0$ unless $x = a$ or $x = b$ where we have, on the other hand, $j_a = -j_b$. If a flow has $j_a = -j_b = 1$, then it is called a *unit flow* from a to b.

Definition 28.2. A *Kirchhoff* flow i in G is a flow with an associated potential function w on V such that

$$i_{xy} = w(x) - w(y) \text{ for all } x, y \text{ which are connected by and edge in } E .$$

Definition 28.3. The energy $E(j)$ of a flow j in G is defined by

$$E(j) = \frac{1}{2} \sum_{x,y} j_{xy}^2 . \tag{28.2}$$

Theorem 28.4 (Thomson's Principle, cf. [DoyS]). *We let i denote a unit Kirchhoff flow in G from vertex a to vertex b and let j denote any other unit flow from a to b. Then*

$$E(i) \le E(j) , \tag{28.3}$$

and equality is obtained if and only if $j = i$.

For the proof we need the following lemma.

Lemma 28.5. *Let w be any given function on V and let d denote a flow in G from a to b. Then*

$$(w_a - w_b)d_a = \frac{1}{2} \sum_{x,y} (w_x - w_y)d_{xy} . \tag{28.4}$$

Proof of the lemma.

$$\sum_{x,y} (w_x - w_y)d_{xy}$$

$$= \sum_x \left(w_x \sum_y d_{xy} \right) - \sum_y \left(w_y \sum_x d_{xy} \right)$$

$$= w_a \sum_y d_{ay} - w_a \sum_x d_{xa} + w_b \sum_y d_{by} - w_b \sum_x d_{xb} \tag{28.5}$$

$$= w_a d_a - w_a(-d_a) + w_b d_b - w_b(-d_b)$$

$$= 2(w_a - w_b)d_a . \qquad \square$$

Proof of Thomson's Principle. We let $d = j - i$ denote the difference flow in G, such that $d_{xy} = j_{xy} - i_{xy}$ for all x and y in V. Then $d_a = d_b = 0$. Let w denote the Kirchhoff potential function on V associated with the Kirchhoff flow i. Then

$$\sum_{x,y} j_{xy}^2 = \sum_{x,y} (i_{xy} + d_{xy})^2$$

$$= \sum_{x,y} i_{xy}^2 + 2 \sum_{x,y} i_{xy}d_{xy} + \sum_{x,y} d_{xy}^2 \tag{28.6}$$

$$= \sum_{x,y} i_{xy}^2 + 2 \sum_{x,y} (w_x - w_y)d_{xy} + \sum_{x,y} d_{xy}^2 ,$$

where the middle term in the last equation is obtained from the fact that the flow i is assumed to be a Kirchhoff flow. This middle term is, however, zero by Lemma 28.5 (because $d_a = 0$), so that

$$\sum_{x,y} j_{xy}^2 = \sum_{x,y} i_{xy}^2 + \sum_{x,y} d_{xy}^2 \ge \sum_{x,y} i_{xy}^2 , \tag{28.7}$$

and [=] is obtained if and only if $d_{xy} = 0$ for all x and y. $\qquad \square$

29. Scherk's Graph is Transient

We use a constructive flow method to show that Scherk's graph is transient in the sense that it allows a unit flow which satisfies the discrete version of the specific Kelvin–Nevanlinna–Royden criterion (15.5). It must be noted, that M. Kanai's approximation theorems (see [Ka1, Ka2, Ka3]) then shows, that Scherk's surface is hyperbolic as well. See the Figures 15–19, which show the explicit construction of the surface and of the corresponding graph.

Theorem 29.1 ([MaMT]). *Scherk's graph G is transient.*

Proof. We show that less than one quarter of the Scherk graph is transient. For this we consider the following subgraph H of G. The vertices of H are the points in $\mathbb{R}^3$ with nonnegative integer coordinates. Two vertices (x, y, z) and (x', y', z') are adjacent if either $(x = x'$ and $|y - y'| + |z - z'| = 1)$ or $(z = z' = 0$ and $|x - x'| + |y - y'| = 1)$. The vertices which have z-coordinate 0 induce a quadrant of $\mathbb{Z}^2$, which we call B. The vertices with a fixed x-coordinate k induce a quadrant of $\mathbb{Z}^2$, which we call H_k. We also let L_k denote the path in H_k with vertices $V(H_k) \cap B$.

 The idea of the proof is to use the vertical 'walls' of the graph H to reorganize a unit flow from the origin to infinity and thereby obtain a unit flow f with finite energy $W(f)$. According to the (discrete version of) Kelvin–Nevanlinna–Royden criterion we then conclude that the graph is transient.

 We first fix two real numbers δ and ϵ such that $0 < 3\epsilon < \delta < 1$.

 In the wall H_n we are only going to use the vertices which have y-coordinates at most $n^{1+3\epsilon}$ and z-coordinate at most n^δ. The active vertices through which we are going to send the flow are therefore all contained in this wedge subgraph of H.

Figure 15.
The building block for Scherk's surface. The surface is the graph surface of the function $\phi(u, v) = \ln\left(\frac{\cos(v)}{\cos(u)}\right)$.

Figure 16.
The checker board positioning of 7 building blocks for Scherk's surface.

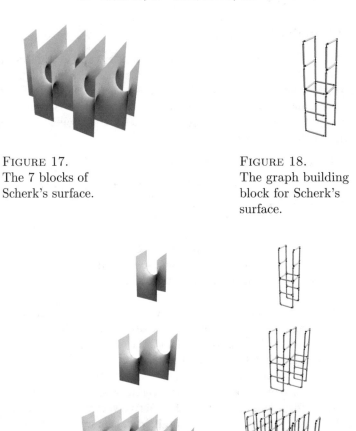

FIGURE 17.
The 7 blocks of
Scherk's surface.

FIGURE 18.
The graph building
block for Scherk's
surface.

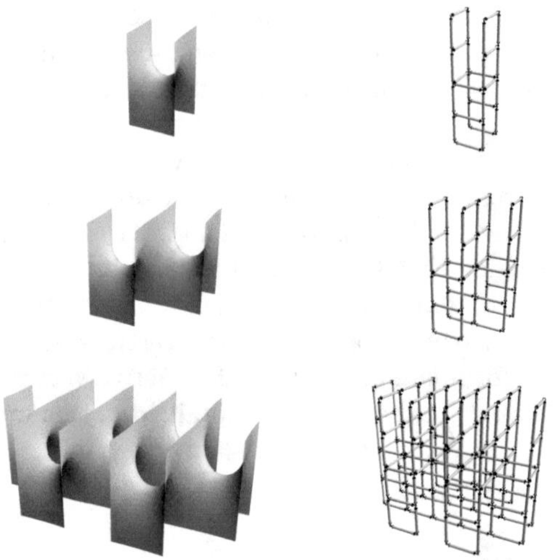

FIGURE 19. The Scherk Buildings – smooth and discrete, respectively.

We are going to construct the flow such that all the flow values in the edges in B which go from L_{n-1} to L_n are at most $n^{-1-2\epsilon}$. This will guarantee, that the contribution to the flow energy $W(f)$ by the edges from L_{n-1} to L_n is at most $\left(n^{1+3\epsilon}\right) \cdot \left(n^{-1-2\epsilon}\right)^2 = n^{-(1+\epsilon)}$. This in turn will guarantee, that the complete contribution from the horizontal edges in H, which are not in some H_n is at most $\sum_n n^{-(1+\epsilon)} < \infty$.

From now on we assume that n is large, and that we have already defined the flow f on the edges in H_i and on the edges from L_i to L_{i+1} for all values of i in $\{0, 1, 2, \ldots, n-1\}$. The task is then to define f on the edges in H_n and on the edges from L_n to L_{n+1} in such a way that the requirements above are satisfied.

Consider a vertex v in L_n. We let $q(v)$ denote the flow that v receives from L_{n-1}. To keep track of whether $q(v)$ is relatively large or relatively small, we define the following auxiliary function

$$\Delta_n = n^{-1-2\epsilon} - (n+1)^{-1-2\epsilon} < 2n^{-2-2\epsilon}, \qquad (29.1)$$

and then we define v to be a *potential receiver* in L_n if $q(v) < (n+1)^{-1-2\epsilon} - \Delta_n$, and v is defined to be a *transmitter* in L_n if $q(v) > (n+1)^{-1-2\epsilon}$. If v is not a transmitter, i.e., $q(v) \le (n+1)^{-1-2\epsilon}$, then we send the flow $q(v)$ directly in the edge from v to L_{n+1}. Otherwise we send some of the flow through that edge, namely the maximally allowed $(n+1)^{-1-2\epsilon}$, and the rest, i.e., the flow $q(v) - (n+1)^{-1-2\epsilon}$, we send to a potential receiver in L_n via the wall H_n as will be explained below.

Firstly we have to guarantee, that there is always a potential receiver available in L_n for each given transmitter:

If v is not a potential receiver, then by definition $q(v) > (n+1)^{-1-2\epsilon} - \Delta_n > \frac{1}{2}n^{-1-2\epsilon}$, and since the net value of the flow from L_{n-1} to L_n is always 1, the number of non-potential receivers is less than $2n^{1+2\epsilon}$. Therefore the number of potential receivers in L_n is at least $n^{1+3\epsilon} - 2n^{1+2\epsilon} > \frac{1}{2}n^{1+3\epsilon}$.

When n is sufficiently large this outnumbers the number of transmitters in L_n: Since the net value of the flow from L_n to L_{n+1} is also 1 the number of transmitters is at most $(n+1)^{1+2\epsilon} < 4n^{1+2\epsilon}$.

We divide the transmitters into $\lceil n^\delta \rceil$ sets such that the number of elements in any two of them differ by at most 1. Then the number of elements in each set is at most $4n^{1+2\epsilon-\delta}$. With each set K we associate precisely one of the $\lceil n^\delta \rceil$ paths, let us call it P_K, in H_n which has a fixed z-coordinate in the allowed range set $\{0, 1, 2, \ldots, \lceil n^\delta \rceil - 1\}$ and which has all its y-coordinates between 0 and $n^{1+3\epsilon}$.

For each transmitter v in K we then choose a potential receiver $r(v)$ and send the rest flow (from above) $q(v) - (n+1)^{-1-2\epsilon}$ from v to P_K (using the shortest path), then through a part of P_K, and then finally through the shortest path from P_K to $r(v)$.

The maximum flow in the shortest paths is at most Δ_n and the maximum flow value in P_K is at most $|K|\Delta_n$. Thus we get an upper bound on the contribution in H_n to $W(f)$ from K, namely

$$2|K|n^\delta \Delta_n^2 + n^{1+3\epsilon}|K|^2 \Delta_n^2 \qquad . \qquad (29.2)$$

Since $|K| \le 4n^{1+2\epsilon-\delta}$ and since there are $\lceil n^\delta \rceil$ such sets K, the total contribution in H_n to $W(f)$ is at most

$$32(n^{-3-2\epsilon+\delta} + 2n^{-1+3\epsilon-\delta}) . \qquad (29.3)$$

Since $3\epsilon < \delta$, we conclude that $W(f)$ is finite and this completes the proof. $\square$

References

[Ahl]   L.V. Ahlfors, *Sur le type d'une surface de Riemann*, C.R. Acad. Sci. Paris, 201, pp. 30–32, (1935).

[AhS]   L.V. Ahlfors and L. Sario, *Riemann surfaces*, Princeton Univ. Press, Princeton, N.J. (1960).

[AHP]   L. Alias, A. Hurtado and V. Palmer, *Geometric Analysis of Lorentzian distance function on spacelike hypersurfaces*, Trans. Amer. Math. Soc., 362, no. 10, pp. 5083–5106, (2010).

[Ana]   C.K. Anand, *Harmonic morphisms of metric graphs*, In 'Harmonic morphisms, harmonic maps, and related topics' (eds.: C.K. Anand, P. Baird, E. Loubeau, and J.C. Wood), Research Notes in Mathematics Vol. 413, Chapman & Hall (1999), 109–112.

[And]   M.T. Anderson, *Complete minimal varieties in hyperbolic space*, Inventiones Math. **69** (1982), 477–494.

[BBC]   R. Banuelos, M. van den Berg, and T. Carroll, *Torsional rigidity and expected lifetime of Brownian motion*, J. London Math. Soc. **66** (2002), 499–512.

[BC]   R. Banuelos and T. Carroll, *Brownian motion and the fundamental frequency of a drum*, Duke Math. J. **17** (1994),575–602.

[B]   J. Barta, *Sur la vibration fondamentale d'une membrane*, C. R. Acad. Sci. **204** (1937), 472–473.

[Ba]   V. Bayle, *Propriétés de concavité du profil isopérimetrique et applications*, PhD thesis, Institut Fourier (Grenoble), 2003.

[BEE]   J.K. Beem, P.E. Ehrlich and K.L. Easey, *Global Lorentzian Geometry*, Marcel Dekker Inc., New York, (1996).

[BenS]   I. Benjamini and O. Schramm, *Random walks and harmonic functions on infinite planar graphs using square tilings*, Annals of Probability **24** (1996), 1219–1238.

[vdB]   M. van den Berg, *Heat Content in Non-compact Riemannian Manifolds*, Integral Equations and Operator Theory, **90** (2018) 90.

[BBV]   M. van den Berg, G. Buttazzo, and B. Velichkov, *Optimization problems involving the first Dirichlet eigenvalue and the torsional rigidity*, New trends in shape optimization, International Series of Numerical Mathematics **166**, Birkhäuser, Springer, Cham (2015) 19–41.

[BFNT]   M. van den Berg, V. Ferone, C. Nitsch, and C. Trombetti, *On Polya's inequality for torsional rigidity and first Dirichlet eigenvalue*, Integral Equations Operator Theory **86** (2016), 579–600.

[BG]   M. van den Berg and P.B. Gilkey, *Heat content and Hardy inequality for complete Riemannian manifolds*, Bull. Lond. Math. Soc. **36** (2004), 577–586.

[BerGM]   M. Berger, P. Gauduchon, and E. Mazet, *Le Spectre d'une Variété Riemannienne*, Lecture Notes in Mathematics **194**, Springer-Verlag (1971).

[BGJ]   [BGJ] G.P. Bessa, V. Gimeno, and L.P. Jorge, *Green functions and the Dirichlet spectrum*, Revista Matemática Iberoamericana, **36** (2020), 1–36.

[BlF]   C. Blanc and F. Fiala, *Le type d'une surface et sa courbure totale*, Comment. Math. Helv. **14** (1941–1942), 230–233.

[CGL]   L. Cadeddu, S. Gallot, and A. Loi, *Maximizing mean exit-time of the Brownian motion on Riemannian manifolds*, Monatshefte für Mathematik **176** (2015), 551–570.

[CKD]   D. Calladay, L. Kaganovskiy, and P. McDonald, *Torsional rigidity, isospectrality and quantum graphs*, J. Phys. A: Math. Theor. **50** (2017), 035201 (15 pp).

[Ca]   P. Cartier, *Fonctions harmoniques sur une arbre*, Symposia Mathematica **9** (1972), 203–270.

[Cha1]   I. Chavel, *Eigenvalues in Riemannian Geometry*, Academic Press (1984).

[Cha2]   I. Chavel, *Riemannian Geometry: A Modern Introduction*, Cambridge Tracts in Mathematics **108**, Cambridge University Press (1993).

[ChaK]   I. Chavel and L. Karp, *Large time behaviour of the heat kernel: the parabolic λ-potential alternative*, Comment. Math. Helvetici **66** (1991), 541–556.

[CheeE]   J. Cheeger and D. Ebin, *Comparison theorems in Riemannian geometry*, North Holland, Amsterdam (1975).

[Che]   R. Chen, *On heat kernel comparison theorems*, J. of Funct. Analysis **165** (1999), 59–79.

[CheLY1] S.Y. Cheng, P. Li, and S.T. Yau, *On the upper estimate of the heat kernel of a complete Riemannian manifold*, Amer. J. Math. **103** (1981), 1021–1063.

[CheLY2] S.Y. Cheng, P. Li, and S.T. Yau, *Heat equations on minimal submanifolds and their applications*, Amer. J. Math. **106** (1984), 1033–1065.

[Chu]   F.R.K. Chung, *Spectral Graph Theory*, CBMS Regional Conference Series in Mathematics **92**, American Mathematical Society (1997).

[Cim]   S. G.-F. Cimmino, *A conjecture on minimal surfaces*, Atti Acc. Lincei Rend. fis. **8**, LXXXII (1988) 639–644.

[CLD]   D. Colladay, J.J. Langford, and P. McDonald, *Comparison results, exit time moments, and eigenvalues on Riemannian manifolds wit a lower Ricci curvature bound*, J. Geom. Analysis **28** (2018), 3906–3927.

[CoG]   T. Coulhon and A. Grigor'yan, *Random walks on graphs with regular volume growth*, Geom. Funct. Anal. **8** (1998), 656–701.

[CHS]   T. Coulhon, I. Holopainen, and L. Saloff-Coste, *Harnack inequality and hyperbolicity for subelliptic p-Laplacians with applications to Picard type theorems*, Geom. Funct. Anal. **11** (2001), 1139–1191; *Erratum*, Geom. Funct. Anal. **12** (2001), 217.

[Da]   E.B. Davies, *Heat kernel bounds, conservation of probability and the Feller property*, J. Anal. Math. **58** (1992), 99–119.

[DGM]   A. Debiard, B. Gaveau, and E. Mazet, *Temps de sortie des boules normales et minoration locale de λ_1*, C. R. Acad. Sci. Paris. Sér. A **278** (1974), 795–798.

[DHKW]   U. Dierkes, S. Hildebrandt, A. Küster, and O. Wohlrab, *Minimal surfaces, Vol. I and II*, Grundlehren der Math. Wiss. 295 and 296, Springer-Verlag (1991).

[Do]   J. Dodziuk, *Difference equations, isoperimetric inequality, and transience of certain random walks*, Trans. Amer. Math. Soc. **284** (1984), 787–794.

[DoKa]   J. Dodziuk and L. Karp, *Spectral and function theory for combinatorial Laplacian*, Contemporary Math. **73** (1988), 25–40.

[DoKe]   J. Dodziuk and W.S. Kendall, *Combinatorial Laplacians and isoperimetric inequality*, In 'From Local Times to Global Geometry, Control and Physics' (ed. K.D. Elworthy), Pitman Research Notes Math. Ser. **150** (1986), pp. 68–74.

[Doy]   P.G. Doyle, *On deciding whether a surface is parabolic or hyperbolic*, Contemporary Mathematics, **73** (1988) 41–48.

[DoyS]   P.G. Doyle and J.L. Snell, *Random walks and electric networks*, The Carus Math. Monographs, Mathematical Association of America, **22** (1984).

[DLD]   E.B. Dryden, J.J. Langford, and P. McDonald, *Exit time moments and eigenvalue estimates*, Bull. Lond. Math. Soc. **49** (2015), 480–490.

[Du]   R.J. Duffin, *The extremal length of a network*, J. Math. Anal. Appl. **5** (1962) 200–215.

[Dy]   E.B. Dynkin, *Markov processes*, Springer Verlag (1965).

[EF]   J. Eells and B. Fuglede, *Harmonic Maps between Riemannian Polyhedra*, Cambridge Tracts in Mathematics, **142**, Cambridge University Press (2001).

[EGK]   F. Erkekoglu, E. Garcia-Rio and D.N. Kupeli, *On level sets of Lorentzian distance function*, General Relativity and Gravitation **35** (2003), 1597–1615.

[EP]   A. Esteve and V. Palmer, *On the characterization of parabolicity and hyperbolicity of submanifolds*, J. Lond. Math. Soc. (2) **84** (2011), no. 1, 120–136.

[FGM]   H. Ferguson, A. Gray, and S. Markvorsen, *Costa's Minimal Surface via Mathematica*, Mathematica in Education and Research **5** (1996), 5–10.

[F]   J.L. Fernandez, *On the existence of Green's function in a Riemannian manifold*, Proceedings of the American Mathematical Society **96** (1986), 284–286.

[Fr]   J. Friedman, *Some geometric aspects of graphs and their eigenfunctions*, Duke Math. J. **69** (1993), 487–525.

[Go]   V. Gol'dshtein and M. Troyanov, *The Kelvin–Nevanlinna–Royden criterion for p-parabolicity*, Math. Z **232** (1999), 607–619.

[Gr]   A. Gray, *Tubes*, Addison-Wesley, Reading (1990).

[GraP]   A. Gray and M.A. Pinsky, *The mean exit time from a small geodesic ball in a Riemannian manifold*, Bull. Sci. Math. (2) **107** (1983), 345–370.

[GreW]   R. Greene and H. Wu, *Function theory on manifolds which possess a pole*, Lecture Notes in Math., vol. 699, Springer-Verlag, Berlin and New York (1979).

[Gri1]   A. Grigor'yan, *Analytic and geometric background of recurrence and non-explosion of the Brownian motion on riemannian manifolds*, Bull. (N.S.) Amer. Math. Soc. **36** (1999), 135–249.

[Gri2]   A. Grigor'yan, *Heat kernels on weighted manifolds and applications*, In *The ubiquitous heat kernel*, volume 398 of *Contem. Math.*, pp. 93–191, Amer. Math. Soc., Providence, RI; International Press, Boston, MA, (2006).

[Gri3]   A. Grigor'yan, *Heat kernel and analysis on manifolds*, volume 47 of *AMS/IP studies in Advanced Mathematics*, Amer. Math. Soc., Providence, RI; International Press, Boston, MA, (2009).

[GriMa]   A. Grigor'yan and J. Masamune, *Parabolicity and stochastic completeness of manifolds in terms of the Green formula*, J. Math. Pures Appl., (9), **100** (2013), no. 5, 607–632.

[GriSa] A. Grigor'yan and L. Saloff-Coste. *Hitting probabilities for Brownian motion on Riemannian manifolds*. J. Math. Pures Appl. (9), **81** (2002), no. 2, 115–142.

[GroKM] D. Gromoll, W. Klingenberg, W. Meyer, *Riemannsche Geometrie im Großen*, Lecture Notes in Mathematics **55**, Springer-Verlag (1975).

[G] M. Gromov, *Isoperimetry of waists and concentration of maps*, Geom. Funct. Anal. **13** (2003), no. 1, 178–215.

[H] R.Z. Ha'sminskii, *Probabilistic representation of the solution of some differential equations*, in Proc. 6th All Union Conf. on Theor. Probability and Math. Statist. (Vilnius 1960), (1960).

[HofM] D. Hoffman and W.H. Meeks III, *The strong halfspace theorem for minimal surfaces*, Invent. Math. **101** (1990), 373–377.

[Hol] I. Holopainen, *Rough isometries and p-harmonic functions with finite Dirichlet integral*, Revista Matematica Iberoamericana **10** (1994), 143–176.

[HPR1] A. Hurtado, V. Palmer, and C. Rosales, *Parabolicity criteria and characterization results for submanifolds of bounded mean curvature in model manifolds with weights*, Nonlinear Anal. **192** (2020), 111681.

[HPR2] A. Hurtado, V. Palmer, and C. Rosales, *Intrinsic and extrinsic comparison results for isoperimetric quotients and capacities in weighted manifolds*, Preprint, arXiv:1907.07920, 2019.

[HMPa] A. Hurtado, S.Markvorsen, and V. Palmer, *Torsional rigidity of submanifolds with controlled geometry*, Math. Ann. **344** (2009), 511–542.

[HMPb] A. Hurtado, S.Markvorsen, and V. Palmer, *Comparison of exit moment spectra for extrinsic metric balls*, Potential Anal. **36** (2012), 137–153.

[HMPc] A. Hurtado, S.Markvorsen, and V. Palmer, *Estimates of the first Dirichlet eigenvalue from exit time moment spectra*, Math. Ann. **365** (2016), 1603–1632.

[I1] K. Ichihara, *Curvature, geodesics and the Brownian motion on a Riemannian manifold I; Recurrence properties*, Nagoya Math. J. **87** (1982), 101–114.

[I2] K. Ichihara, *Curvature, geodesics and the Brownian motion on a Riemannian manifold II; Explosion properties*, Nagoya Math. J. **87** (1982), 115–125.

[I3] K. Ichihara, *Comparison theorems for Brownian motions on Riemannian manifolds and their applications*, Journal of Multivariate Analysis **24** (1988), 177–188.

[JK] L.P. Jorge and D. Koutroufiotis, *An estimate for the curvature of bounded submanifolds*, Amer. J. Math. **103** (1981), 711–725.

[J] J. Jost, *Partial differential equations*, Springer (2013).

[Kac] M. Kac, *Can one hear the shape of a drum?*, Amer. Math. Monthly **73** (1966), 1–23.

[K1] S.L. Kalpazidou, *On Beurling inequality in terms of thermal power*, J. Appl. Prob. **28** (1991), 104–115.

[K2] S.L. Kalpazidou, *Cycle representations of Markov processes*, Springer-Verlag (1995).

[Ka1] M. Kanai, *Rough isometries and combinatorial approximations of geometries of non-compact Riemannian manifolds*, J. Math. Soc. Japan **37** (1985), 391–413.

[Ka2]   M. Kanai, *Rough isometries and the parabolicity of Riemannian manifolds*, J. Math. Soc. Japan **38** (1986), 227–238.

[Ka3]   M. Kanai, *Analytic inequalities and rough isometries between non-compact Riemannian manifolds*, In 'Curvature and topology of Riemannian manifolds' (eds. K. Shiohama, T. Sakai, and T. Sunada), Lecture Notes in Mathematics 1201 (Springer, Berlin, 1986), pp. 122–137.

[Kar]   H. Karcher, *Riemannian comparison constructions*. In 'Global Differential Geometry' (Ed. S.S. Chern), MAA Studies in Mathematics **27** (1989), 170–222.

[KarP]   L. Karp and M. Pinsky, *Volume of a small extrinsic ball in a submanifold*, Bull. London Math. Soc. **21** (1989), 87–92.

[KD]   K.K. Kinateder and P. McDonald, *Variational principles for average exit time moments for diffusions in Euclidean space*, Proc. Amer. Math. Soc. **127** (1999), 2767–2772.

[KDM]   K.K. Kinateder, P. McDonald, and D. Miller, *Exit time moment, boundary value problem, and the geometry of domains in Euclidean space*, Probab. Theory Related Fields **111** (1998), 469–487.

[KR]   D.J. Klein and M. Randić, *Resistance distance*, J. Math. Chem. **12** (1993), 81–95.

[Li]   P. Li, *Curvature and function theory on Riemannian manifolds*, Surveys in Differential Geometry "In Honor of Atiyah, Bott, Hirzebruch, and Singer", vol. VII, International Press, Cambridge, 2000, pp. 71–111.

[LiT]   P. Li and L.F. Tam, *Symmetric Green's functions on complete manifolds*, Amer. J. Math. **109** (1987), 1129–1154.

[LiW]   P. Li and J. Wang, *Mean value inequalities*, Indiana Univ. J. **48** (1999), 1257–1283.

[Ly.R]   R. Lyons, *Diffusions and random shadows in negatively-curved manifolds* J. Functional Analysis **138** (1996), 426–448.

[Ly.T]   T. Lyons, *A simple criterion for transience of a reversible Markov chain*, Ann. Probab. **11** (1983), 393–402.

[L]   J. Lott, *Some geometric properties of the Bakry–Emery–Ricci tensor*, Comment. Math. Helv. **78** (2003), no. 4, 865–883.

[LS]   T. Lyons and D. Sullivan, *Function theory, random paths, and covering spaces*, J. Differential Geom. **19** (1984), 299–323.

[MRS]   L. Mari, M. rigoli, and A.G. Setti, *Keller–Osserman conditions for diffusion-type operators on Riemannian manifolds*, J. Funct. anal. **258** (2010), no. 2, 665–712.

[Ma1]   S. Markvorsen, *A sufficient condition for a compact immersion to be spherical*, Math. Z. **183** (1983), 407–411.

[Ma2]   S. Markvorsen, *On the bass note of compact minimal immersions*, Preprint, Max Planck Institut für Mathematik, Bonn (1985), 26 pages.

[Ma3]   S. Markvorsen, *On the heat kernel comparison theorems for minimal submanifolds*, Proc. Amer. Math. Soc. **97** (1986), 479–482.

[Ma4]   S. Markvorsen, *A transplantation of heat kernels and eigenfunctions via harmonic maps*, In Proceedings of the VIth Int. Coll. on Differential Geometry,

Santiago (Spain), Sept. 1988 (Ed. L.A. Cordero), Universidade de Santiago de Compostela (1989).

[Ma5] S. Markvorsen, *A characteristic eigenfunction for minimal submanifolds*, Math. Z. **202** (1989), 375–382.

[Ma6] S. Markvorsen, *On the mean exit time from a minimal submanifold*, J. Diff. Geom. **29** (1989), 1–8.

[Ma7] S. Markvorsen, *Minimal Webs in Riemannian Manifolds*; Geometriae Dedicata **133** (2008), 7–34.

[MaMT] S. Markvorsen, S. McGuinness, and C. Thomassen, *Transient random walks on graphs and metric spaces with applications to hyperbolic surfaces*, Proc. Lond. Math. Soc. **64** (1992), 1–20.

[MaP1] S. Markvorsen and V. Palmer, *Generalized isoperimetric inequalities for extrinsic balls in minimal submanifolds*, J. Reine Angew. Math. **551** (2002), 101–121.

[MaP2] S. Markvorsen and V. Palmer, *The relative volume growth of minimal submanifolds*, Archiv der Mathematik **79** (2002), 507–514.

[MaP3] S. Markvorsen and V. Palmer, *On the isoperimetric rigidity of extrinsic minimal balls*, J. Differential Geom. Appl. **18** (2003), 47–54.

[MaP4] S. Markvorsen and V. Palmer, *Transience and capacity of minimal submanifolds*, Geom. Funct. Anal. **13** (2003), no. 4, 915–933.

[MaP5] S. Markvorsen and V. Palmer, *How to obtain transience from bounded radial mean curvature*, Trans. amer. Math. Soc. **357** (2005), no. 9, 3459–3479.

[MaP6] S. Markvorsen and V. Palmer, *Torsional rigidity of minimal submanifolds*, Proc. London Math. Soc. (3) **93** (2006), no. 1, 253–272.

[MaP7] S. Markvorsen and V. Palmer, *Extrinsic isoperimetric analysis of submanifolds with curvatures bounded from below*, J. Geom. Anal. **20** (2010), no. 2, 388–421.

[McDa] P. McDonald, *Isoperimetric conditions, Poisson problems, and diffusions in Riemannian manifolds*, Potential Analysis **16** (2002), 115–138.

[McDb] P. McDonald, *Exit times, moment problems and comparison theorems*, Potential Analysis **38** (2013), 1365–1372.

[McDMa] P. McDonald and R. Meyers, *Diffusions on graphs, Poisson problems and spectral geometry*, Transactions of the AMS. **354** (2002), 5111–5136.

[McDMb] P. McDonald and R. Meyers, *Isospectral polygons, planar graphs and heat content*, Proceedings of the AMS. **131** (2003), 3589–3599.

[McDM] P. McDonald and R. Meyers, *Dirichlet spectrum and heat content*, J. Funct. Anal. **200** (2003), 150–159.

[Mi] J. Milnor, *On deciding whether a surface is parabolic or hyperbolic*, American Math. Monthly **84** (1977), 43–46.

[Mo] F. Morgan, *Geometric measure theory. A beginner's guide*, Elsevier/Academic Press, Amsterdam, Fourth edition, (2009).

[Mo2] F. Morgan, *Myers' theorem with density*, Kodai Math. J. **29** (2006), no. 3, 455–461.

[O'N] B. O'Neill, *Semi-Riemannian Geometry; With Applications to Relativity*, Academic Press (1983).

[OU]   Y. Ohno and H. Urakawa, *On the first eigenvalue of the combinatorial Laplacian for a graph*, Interdiscip. Inform. Sci. **1** (1994), 33–46.

[P]   J.L. Palacios, *Resistance distance in graphs and random walks*, Int. J. of Quantum Chemistry **81** (2001), 29–33.

[Pa]   V. Palmer, *Isoperimetric Inequalities for extrinsic balls in minimal submanifolds and their applications*, Jour. London Math. Soc. **60**, (2) (1999), 607–616.

[Pa1]   V. Palmer, *On deciding whether a submanifold is parabolic of hyperbolic using its mean curvature*, Simon Stevin Institute for Geometry, Tilburg, The Netherlands, 2010. Simon Stevin Transactions on Geometry, vol 1.

[Pe]   P. Petersen, *Riemannian geometry*, volume 171 of Graduate Texts in Mathematics. Springer, New York, second edition, 2006.

[PW]   M. Picardello and W. Woess (editors), *Random walks and discrete potential theory*, Symposia Mathematica Vol. XXXIX, Cambridge University Press (1999).

[PRRS]   S. Pigola, M. Rigoli, M. Rimoldi, and A. G. Setti, *Ricci almost solitons*, Ann. Sc. Norm. Super. Pisa Cl. Sci. (5) **10** (2011), no. 5, 757–799.

[PRS]   S. Pigola, M. Rigoli, and A.G. Setti, *Vanishing and finiteness results in geometric analysis*, volume 266 of Progress in Mathematics. Birkhäuser Verlag, Basel, 2008. A generalization of the Bochner technique.

[Pin]   M. Pinsky, *Brownian motion, exit times and stochastic Riemannian geometry*, Mathematics and Computers in Simulation **XXVI**, (1984), 357–360.

[Po]   G. Polya, *Über eine Aufgabe der Wahrscheinlichkeitstheorie betreffend die Irrfahrt im Straßennetz*, Math. Ann. **84**, (1921), 149–160.

[Po2]   G. Pólya, *Torsional rigidity, principal frequency, electrostatic capacity and symmetrization*, Quart. Appl. Math. **6**, (1948), 267–277.

[PS]   G. Pólya and G. Szegö, *Isoperimetric Inequalities in Mathematical Physics*, Princeton University Press (1951).

[Q]   Z. Qian *Estimates for weighted volumes and applications*, Quart. J. Math. Oxford Ser. (2) **48** (1997), no. 190, 235–242.

[Re]   D. Revuz, *Markov Chains*, North Holland Mathematical Library, Vol. 11 (1975).

[Sa]   T. Sakai, *Riemannian Geometry*, Transl. Math. Monographs **149**, American Mathematical Society (1999).

[So]   P.M. Soardi, *Potential Theory on Infinite Networks*, Lecture Notes in Mathematics **1590**, Springer-Verlag (1999).

[T1]   C. Thomassen, *Resistances and currents in infinite electrical networks*, J. Combin. Theory Ser. B **49** (1990), 87–102.

[T2]   C. Thomassen, *Isoperimetric inequalities and transient random walks on graphs*, Ann. Probab. **20** (1992), 1592–1600.

[Ur1]   H. Urakawa, *Heat kernel and Green kernel comparison theorems for infinite graphs*, J. Functional Analysis, **146** (1997), 206–235.

[Ur2]   H. Urakawa, *Eigenvalue comparison theorems of the discrete Laplacians for a graph*, Geometriae Dedicata, **74** (1999), 95–112.

[Ur3]   H. Urakawa, *A discrete analogue of the harmonic morphism and Green kernel comparison theorems*, Glasgow Math. J. **42** (2000), 319–334.

[Ur4]   H. Urakawa, *A discrete analogue of the harmonic morphism*, In "Harmonic morphisms, harmonic maps, and related topics" (eds.: C.K. Anand, P. Baird, E. Loubeau, and J.C. Wood), Research Notes in Mathematics Vol. 413, Chapman & Hall (1999), 97–108.

[Wa]   S. Wagon, *The Costa surface, in snow and in Mathematica*, Mathematica in Education and Research **8** (1999), 56–63.

[WW]   G. Wei and W. Wylie, *Comparison geometry for the Bakry–Emery Ricci tensor*, J. Differential Geom. **83** (2009), no. 2, 377–405.

[Wo]   W. Woess, *Random walks on infinite graphs and groups*, Cambridge Tracts in Mathematics **138**, Cambridge University Press (2000).

[W]   W. Wylie, *Sectional curvature for Riemannian manifolds with density*, Geom. Dedicata **178** (2015), 151–169.

[Zhu]   S. Zhu, *The comparison geometry of Ricci curvature* In Comparison geometry (Berkeley, CA, 1993–94), volume 30 of Math. Sci. Res. Inst. Publ., pages 221–262. Cambridge Univ. Press, Cambridge, 1997.

The Dirac Operator in Geometry and Physics

Maung Min-Oo

Foreword

The theme of these notes is centered around the use of the Dirac operator in geometry and physics, with the main focus on scalar curvature, Gromov's K-area and positive mass theorems in General Relativity. The notes are written in a rather impressionistic style following Hermann Weyl's advice. I quote: "The stringent precision attainable in Mathematics has led many authors to a mode of writing which must give the reader the impression of being shut up in a brightly illuminated cell where every detail sticks out with the same dazzling intensity, but without relief. I prefer the open landscape with a clear sky with its depth of perspective where a wealth of sharply defined nearby details gradually fade away towards the horizon".

My objective is to give a flavour of some selected aspects of the subject that I understand, rather than a comprehensive survey. I have not stated results and proofs in their most general form or in their most recent version, to avoid getting bogged down in too many technical details. Instead, I have attempted to give a broader overview of the topics at an introductory level, emphasizing more the basic ideas and concepts involved, so that the reader can more easily consult the literature. Although I have included a number of important references to the original literature, my list is certainly not complete and I apologize for any serious omissions, but nowadays, it is a simple matter to access recent literature through preprint servers, such as the ArXiv, and the MathSciNet.

Foreword to the Second Edition

Except for adding a few sentences and a few results here and there to clarify certain statements, correcting some minor typos and updating some references, I did not change too much to the main body of the text from the first edition of these notes, in order to keep the original spontaneous spirit of the lectures I gave. However, I have added an epilogue where I took the liberty to describe and interpret, from my point of view, some new developments by other researchers that are related to the topics of these lectures.

© Springer Nature Switzerland AG 2020

A. Hurtado et al., *Global Riemannian Geometry: Curvature and Topology*, Advanced Courses in Mathematics - CRM Barcelona, https://doi.org/10.1007/978-3-030-55293-0_2

1. Spinors and the Dirac Operator

The Dirac equation was first introduced by the physicist P.A.M. Dirac in 1928 to describe spin-$\frac{1}{2}$ fermions, such as electrons, in quantum mechanics, incorporating special-relativistic invariance. The equation predicted anti-matter, such as positrons. The mathematical definition (and generalization) of the Dirac operator and spinors on differentiable manifolds, valid in all dimensions, was then given later by Atiyah, Bott and Singer in their quest to prove the celebrated Index Theorem for elliptic operators. The Dirac operator is a first-order differential operator acting on spinorial objects rather than on scalars or differential forms like the usual Laplace operator Δ. However it is, in a precise sense to be described below, a square root of Δ. It is the fundamental operator that occurs in the proof of the Index Theorem. The main role that the Dirac operator plays in Differential Geometry is through its interaction with scalar curvature as was first discovered by A. Lichnerowicz. Scalar curvature and Ricci curvatures play a fundamental role in General Relativity. The duality between fermions and bosons is a key ingredient of modern theoretical physics, such as in super-gravity and super-symmetry. In fact, the most insightful proof of the Index Theorem, is the one given by physicists, such as Witten, using the heat kernel of the square of the Dirac operator (propagator) and interpreting the index as the supersymmetric trace (Witten index).

1.1. Introduction to Spinors

Spinors are objects that are more sensitive to the action of the orthogonal group than ordinary vectors. The simplest example is the Hopf bundle over S^2. This bundle is half as curved as the tangent bundle of S^2 geometrically and topologically. Parallel translation in this bundle around a great circle in S^2 rotates a spinor by an angle of π instead of 2π for a vector and the Euler characteristic of the Hopf bundle is 1, which is half that of the tangent bundle. More generally, the existence of such a double cover of the orthonormal frame bundle, called a spin structure, would be guaranteed by the vanishing of a suitable characteristic class, namely the second Stiefel–Whitney class.

There are two steps involved in defining a spin structure on a manifold. The first is a purely algebraic construction of Clifford algebras and spin groups associated to a vector space with a given quadratic form (scalar product). The second has to do with the way these structures can be globally defined on the tangent bundle or more generally on vector bundles. A rough definition of a spin structure is therefore an assignment of a spin group and spinors at each point on the manifold in a manner that is consistent with the transition functions of the tangent bundle.

We will begin with a brief description of Clifford algebras, which are the devices needed to give an explicit construction of the spin groups. After that, we will turn to the problem of globalizing such objects to manifolds, i.e., to the problem of defining a spin structure on a smooth manifold. It is at this stage that global topological features of the tangent bundle become important. Before turning to the algebraic part, it is useful to mention that 'two step' constructions from an algebraic infinitesimal object to a global definition are quite canonical in differential geometry. A metric on a manifold, for example, is a smoothly varying choice of inner product on the tangent space at each point of the manifold. The construction thus involves extending inner products on each tangent space smoothly to the tangent bundle. For positive definite inner products this extension is always possible. On the other hand, to define an orientation on a manifold, we must consistently extend the orientation defined for each tangent space to the whole manifold and this is in general, not possible. There is a global obstruction to this is called the first Stiefel–Whitney class. Similarly, for the existence of a spin structure on an oriented manifold the obstruction is the second Stiefel–Whitney class. Now for the algebra:

A *quadratic space* is a pair (V, q), where V is a real finite-dimensional vector space (over $\mathbb{K} = \mathbb{C}$ or $\mathbb{R}$) and $q : V \otimes V \to \mathbb{R}$ is a non-degenerate (not necessarily positive-definite) symmetric bilinear form. We denote by $O(V, q)$ the orthogonal group of q. The real *Clifford algebra* $Cl(V, q)$ is the (unique) unital associative algebra generated by V subject to the defining relation

$$v \cdot w + w \cdot v = -2q(v, w)1$$

for every $v, w \in V$.

V is naturally imbedded in $Cl(V, q)$ and if c is any linear map from V to a unital algebra satisfying $c(v) = -q(v, v)$, then c extends naturally to an algebra homomorphism from $Cl(V, q)$ and the Clifford algebra is the universal algebra with this characterizing property. As a vector space the Clifford algebra is isomorphic to the exterior algebra $\bigwedge V$ of V, but the multiplication is, of course, different. In fact, $\bigwedge V$ has a natural module structure over the Clifford algebra defined by extending the map: $V \to End(\bigwedge V)$, $v \mapsto \gamma(v) = v \wedge + v \lrcorner$ where $\lrcorner$ denotes interior multiplication which is the transpose of the exterior multiplication $\wedge$.

It can also be described as the algebra $Cl(V, q) = \mathcal{T}(V)/\mathcal{I}$. Here $\mathcal{T}(V)$ is the tensor algebra $\mathcal{T}(V) = \mathbb{K} \oplus V \oplus (V \otimes V) \oplus \cdots$ and $\mathcal{I}$ is the ideal generated by all elements of the form $v \otimes w + w \otimes v + 2q(v, w)1$, where 1 is the unit element of the tensor algebra. V injects as a subspace in $Cl(V, q)$ since the ideal $\mathcal{I}$ is disjoint from V and any orthonormal basis of V generates the Clifford algebra.

The map $Cl(V, q) \to \bigwedge V$, $v \mapsto \gamma(v)1$, where 1 is the unit in $\bigwedge^0 V \simeq \mathbb{K}$, is a vector space isomorphism.

The isometry $v \mapsto -v$ extends to an involutive automorphism of the algebra defining its $\mathbb{Z}/2$-grading:

$$Cl(q) = Cl^0(q) \oplus Cl^1(q).$$

If $V = \mathbb{R}^m$ and q is the standard positive definite quadratic form, then we will simply write $Cl_m = Cl^0_m \oplus Cl^1_m$.

The map $\mathbb{R}^m \to Cl^0_{m+1}$, $v \mapsto ve_{m+1}$, extends to an isomorphism of algebras: $Cl_m \cong Cl^0_{m+1}$.

If $(e_1, \ldots, e_m)$ is an orthonormal frame in V, then $\eta = e_1 \cdots e_m \in Cl(q)$ is called the *volume element*. The square of η is either 1 or -1, depending on the signature of q.

Let (M, g) be a Riemannian manifold. The *Clifford bundle* of (M, g) is the total space $Cl(M, g) = \bigcup_x Cl(T_xM, g_x)$ of all the Clifford algebras of the tangent spaces.

A bundle of *Clifford modules* on (M, g) is a complex vector bundle S over M with a homomorphism of bundles of algebras $\gamma : Cl(M, g) \to End(S)$, i.e., for every $x \in M$, the vector space S_x is a left module over the algebra $Cl(T_xM, g_x)$. Restricted to $TM \subset Cl(g)$, the map γ is a *Clifford morphism*, i.e., a homomorphism of vector bundles such that $\gamma(v)^2 = -|v|^2 id_{S_x}$ for every $x \in M$ and $v \in T_xM$.

It follows from the universal property of Clifford algebras that conversely, given a vector bundle S over M and a Clifford morphism: $TM \to End(S)$, one can extend it to a homomorphism of bundles of algebras.

Here are two examples:

(i) The bundle of *exterior algebra* on M. Put $S = \bigwedge T^*M$ and define γ by $\gamma(v)\omega = v \lrcorner \omega + v \wedge \omega$ for $v \in T_xM$ and $\omega \in S_x$ where $\lrcorner$ denotes interior multiplication with respect to the given metric.

   The isomorphism $Cl(M, g) \cong \bigwedge T^*(M)$ given by the map $v \mapsto \gamma(v) \cdot 1$ is called the symbol map. It is the symbol of the de Rham operator $d + \delta$ on the exterior algebra of forms.

(ii) Let (M, g) be a *Kähler manifold* with complex structure J. The map: $\gamma(v + \bar{v})\omega = \sqrt{2}(\bar{v} \lrcorner \omega + v \wedge \omega)$ for $w \in W = \{v \in TM \otimes \mathbb{C} \,|\, J(v) = iv\}$ and put $\omega \in S = \bigwedge W$ defines a Clifford module.

If $n \geq 3$ then the fundamental group of the special orthogonal group $SO(n)$ is $\mathbb{Z}/2$ and the simply connected universal cover is a group called $\mathrm{Spin}(n)$. We will use Clifford algebras to describe this group.

First of all, the pinor group $\mathrm{Pin}(m)$ is the group consisting of all products of unit vectors in Cl_n. The map $\rho(v) : x \mapsto v \cdot x \cdot v^{-1}$, where v is a unit vector and x is any vector in $\mathbb{R}^m$, describes the reflection in the hyperplane $v^\perp$ and hence defines a representation from $\mathrm{Pin}(m)$ to $O(m)$ which is a double cover. Since $O(m)$ has two connected components, we can restrict to the pre-image of the identity component $SO(m)$ to obtain the spinor group $\mathrm{Spin}(m)$. Therefore $\mathrm{Spin}(m) = \mathrm{Pin}(m) \cap Cl_m^0$, and $\mathrm{Spin}(m)$ is generated by products of an even number of unit vectors in the Clifford algebra. In fact:

$$\mathrm{Spin}(m) = \{v_1 \cdots v_{2l} \in Cl_m \,|\, q(v_i, v_i) = \pm 1\}.$$

We can also complexify the Clifford algebra $Cl_m^c = Cl_m \otimes \mathbb{C}$ and define the complex spinor group as $\mathrm{Spin}^c(m) = \mathrm{Spin}(m) \otimes_{\mathbb{Z}/2} S^1$.

One basic property of the spinor group is that there exists (half-integral) representations which do not descend to $SO(n)$. The basic representation space is called the space of spinors and these are the "sensitive" objects, since the orthogonal group cannot act as a single-valued representation.

In even dimensions $m = 2n$, the algebra Cl_m is a simple matrix algebra, and there is a unique faithful and irreducible (graded) Dirac representation in a complex, 2^n-dimensional vector space $\mathbb{S}$ called the spinor space such that $Cl_m \otimes \mathbb{C} = End(\mathbb{S})$. Restricted to Cl_m^0 (and hence to $Spin(n)$), this representation decomposes into the direct sum of two irreducible and inequivalent, half-spinor Weyl representations $\mathbb{S} = \mathbb{S}^+ \oplus \mathbb{S}^-$. The splitting is basically given by the eigenspaces of Clifford multiplication with the volume element η. In odd dimensions, we can use the isomorphism: $Cl_{2n} \cong Cl_{2n+1}^0$ to obtain the unique irreducible complex spinor representation of dimension 2^n. (The representation does not split w.r.t. $Spin(2n+1)$). There are exactly two irreducible representations of Cl_{2n+1}, of complex dimension 2^n which become isomorphic representations when restricted to $Spin(2n+1)$, since the volume form η is now the intertwining map.

The *Stiefel–Whitney classes* $w(E)$ of a real vector bundle E of rank r are $\mathbb{Z}/2$-characteristic classes characterized by the following properties:

(i) $w(E) = 1 + w_1(E) + \cdots + w_r(E)$ with $w_i(E) \in H^i(M; \mathbb{Z}/2)$.
(ii) If $f : M \to N$ is a map then $f^*(w(E)) = w(f^*(E))$.
(iii) $w(E \oplus F) = w(E) + w(F)$.
(iv) If L is the non orientable Möbius line bundle over S^1 then $w(L) \neq 0$.

The first Stiefel–Whitney class measures the obstruction for a vector bundle to be orientable and the non-vanishing of the second is the obstruction to the existence of a spin structure. Stiefel–Whitney classes, like Chern and Pontryagin classes are stable in the sense that they depend only on the stable class of the vector bundle, i.e., $w(E \oplus k_\mathbb{R}) = w(E)$ where $k_\mathbb{R}$ denotes the trivial (real) vector bundle of dimension k.

Let E be an oriented vector bundle with a fiber metric over a manifold M and let U_α be a simple cover of M such that E has transition functions $g_{\alpha\beta} \in SO(r)$ on $U_\alpha \cap U_\beta$ satisfying the cocycle condition $g_{\alpha\beta}g_{\beta\gamma} = g_{\alpha\gamma}$.

We say that E admits a spin structure if E is orientable and we can define lifts $\tilde{g}_{\alpha\beta}$ of the transition functions to $Spin(r)$ such that the cocycle condition is preserved. This can be expressed in terms of the Stiefel–Whitney classes simply as $w_2(E) = 0$. The set of all inequivalent spin structure is then parametrized by $H^1(M; \mathbb{Z}/2)$. We say that a Riemannian manifold is spin if the tangent bundle TM admits a spin structure.

From the point of view of principal bundles, a spin structure can be realized as a principal bundle $Spin(E)$ with structure group $Spin(n)$ which is a double cover (over each fiber) of the oriented frame bundle $SO(E)$ of the vector bundle E. Given a spin Riemannian manifold, the vector bundle associated to the basic spinor representation will be denoted by $\mathbb{S}$ and is called the spinor bundle.

Similarly, the necessary and sufficient topological condition to define a $Spin^c$ structure on a unitary bundle E is that $w_2(E)$ is the mod 2 reduction of an integral cohomology class. This is always true for a Hermitian vector bundle E, since $w_2(E) \equiv c_1(E) \mod 2$.

Examples

1. The sphere S^n is spin for all n, since the tangent bundle is stably trivial. In fact, $T(S^n) \oplus 1$ is trivial.
2. The complex projective space is spin if and only if m is odd, since the first Chern class is given by $m + 1$ times the generator in $H^2(\mathbb{C}P^m; \mathbb{Z})$. In fact, $T(\mathbb{C}P^m) \cong H \otimes H^{\perp}$ where H is the canonical line bundle and $H^{\perp}$ the complementary m-plane with $H \oplus H^{\perp}$ trivial, so that:

$$T(\mathbb{C}P^m) \oplus 1_{\mathbb{C}} \cong (H^* \otimes H^{\perp}) \oplus (H^* \otimes H) \cong (m + 1)H^*.$$

However, as is true for all Kähler manifolds, $\mathbb{C}P^m$ has a canonical Spin^c structure for all m. The twisting line bundle is H, since

$$H \otimes (T(\mathbb{C}P^m) \oplus 1) \cong (m + 1)(H^* \otimes H) \cong (m + 1)$$

is trivial. This also shows that $c_1(\mathbb{C}P^m) = (m + 1)c_1(H^*) = m + 1$ which is even when m is odd.
3. All Kähler manifolds with even first Chern class are spin.
4. All parallelizable manifolds, in particular Lie groups, are spin.
5. The product and connected sum of spin manifolds is again spin.

1.2. The Dirac Operator

The Dirac operator is the fundamental first-order elliptic operator defined on a spin manifold. Its symbol is given by Clifford multiplication and its index is given by the $\hat{A}$-genus of the spin manifold.

Let M^n be a Riemannian manifold with Clifford bundle $Cl(M)$ and let S be any bundle of left modules over $Cl(M)$. Assume that S is furnished with a metric and a connection preserving the metric and compatible with the Clifford module structure, i.e., the product rules hold:

$$\nabla \langle s_1, s_2 \rangle = \langle \nabla s_1, s_2 \rangle + \langle s_1, \nabla s_2 \rangle$$

for any two sections $s_1, s_2 \in \Gamma(S)$ and

$$\nabla(w \cdot s) = \nabla w \cdot s + w \cdot \nabla s$$

for any $w \in Cl(M)$ and $s \in \Gamma(S)$.

Then the Dirac operator of S is the canonical first-order differential operator defined by:

$$D\sigma = \sum_{k=1}^{n} e_k \cdot \nabla_{e_k} \sigma$$

where $\{e_k\}$ is an orthonormal base of TM and $\sigma \in \Gamma(S)$. To see that it is a globally well-defined operator independent of the choice of the orthonormal base we can express D as a composition:

$$\Gamma(\mathbb{S}) \to \Gamma(T^*(M) \otimes \mathbb{S}) \to \Gamma(\mathbb{S})$$

where the first map is the covariant derivative ∇ and the second map is Clifford (co-)multiplication.

This defines a general Dirac operator since all we need is a Clifford module over a (not necessarily spin) manifold. However, we will be mainly interested in twisted Dirac operators that are defined on $\mathbb{S} \otimes E$ where E is a complex Hermitian bundle with a connection and $\mathbb{S}$ is the spinor bundle. These operators can sometimes be defined even if the manifold is not spin, provided the tensor product exists as a bundle. For example for Kähler manifolds we always have the *spin^c* Dirac operator where E is a line bundle which can be thought of as the "virtual square root" of the canonical bundle. Although E and $\mathbb{S}$ do not exist globally on the manifold, $\mathbb{S} \otimes E$ and $E \otimes E$ are well defined.

For even-dimensional manifolds the spinor representation has a natural splitting $\mathbb{S} = \mathbb{S}^+ \oplus \mathbb{S}^-$ and the Dirac operator splits as $D = D^+ + D^-$ with $D^\pm$: $\mathbb{S}^\pm \otimes E \longrightarrow \mathbb{S}^\mp \otimes E$, and D^- is the adjoint of D^+.

Since the Dirac operator on a closed compact manifold is a self-adjoint elliptic operator it has a real discrete spectrum with finite multiplicities on a compact manifold. In particular, the index of D^+:

$$\text{index}(D^+) = \dim(\text{Kern}(D^+)) - \dim(\text{Kern}(D^-))$$

is a topological invariant given by the famous Atiyah–Singer Index Theorem:

$$\text{index}(D^+) = \int_M \hat{A}(M) \wedge ch(E)$$

where the $\hat{A}$ genus, a certain formal power series in the Pontryagin classes of M, and $ch(E)$, the Chern character of the vector bundle E, will be defined below.

The Chern character of a complex bundle E of rank r can be defined by

$$ch(E) = \sum_{k=1}^{r} \exp(x_k)$$

where the total Chern class is expressed (by the splitting principle) as:

$$C(E) = 1 + c_1(E) + \cdots + c_r(E) = \prod_{k=1}^{r}(1 + x_k),$$

so that c_k is given by the kth elementary symmetric function of the x_k's. The first few terms are:

$$ch(E) = \dim(E) + c_1(E) + \frac{1}{2}(c_1(E)^2 - 2c_2(E)) + \cdots$$

The Chern character satisfies:

$$ch(E_1 \oplus E_2) = ch(E_1) + ch(E_2), \quad ch(E_1 \otimes E_2) = ch(E_1)ch(E_2)$$

and hence defines a ring homomorphism $ch : K(M) \longrightarrow H^{\text{even}}(M)$.

Similarly, the total $\hat{A}$ genus is given by:

$$\hat{A}(M) = \prod_{k=1}^{r} \frac{x_k/2}{\sinh(x_k/2)}$$

where now the total Pontryagin class of TM is formally expressed as:

$$p(M) = 1 + p_1(M) + \cdots + p_r(M) = \prod_{k=1}^{r}(1 + x_k^2),$$

so that p_k is given by the kth elementary symmetric function of the x_k^2's. The first few terms are:

$$\hat{A} = 1 - \frac{1}{24}p_1 + \frac{1}{2^7 3^2 5}(-4p_2 + 7p_1^2) + \cdots$$

We can represent the Chern character of E by the differential form

$$ch(E) = Tr\left(\exp\left(\frac{F^{\nabla}}{2\pi i}\right)\right)$$

where F^{∇} is the curvature of a connection ∇ for E, regarded as an $End(E)$-valued two form.

Similarly, $\hat{A}(M)$ is represented by the closed differential form:

$$\hat{A}(M) = \sqrt{\det}\left(\frac{R/2}{\sinh(R/2)}\right)$$

where R is the Riemannian curvature of the metric g, regarded as an $End(TM)$-valued two form, and $\sqrt{\det}$ is the Pfaffian, which is an invariant polynomial defined on the Lie algebra of skew symmetric matrices in even dimensions.

The modern "supersymmetric" proof of the index theorem (see [BGV]) is based on the asymptotics of the (super-)trace of the heat kernel $\exp(-tD^2)$ as $t \to 0^+$ and as $t \to \infty$. The local index density $\hat{A}(M)ch(E)$ term is obtained in the limit as $t \to 0^+$ evaluated on the diagonal in $M \times M$ and the global harmonic projectors appear as $t \to \infty$ after things have settled down harmoniously. This is another example of the transition from local to global invariants that lies at the heart of modern differential geometry. Since the supertrace of the heat kernel is independent of t, one obtains an expression for the analytical index, which is the difference between the dimensions of the harmonic spinors for D^+ and D^- in terms of characteristic classes, which in turn, can be obtained by integrating the top differential form (fermionic integration).

1.3. The Lichnerowicz Formula

The Bochner technique of proving vanishing theorems for harmonic forms rely on expressing the relevant Laplacian as a sum of a non-negative differential operator (the rough Laplacian) and a purely algebraic term depending only curvature. This is also known as a Weitzenböck formula. It is a powerful technique, since one gets a global result, such as non-existence of harmonic objects, by assuming appropriate local curvature assumptions, connecting global topology to local geometry. One year after the publication of the Atiyah–Singer Index Theorem, A. Lichnerowicz, in 1963, proved his now famous formula for the square of the Dirac operator:

$$D^2 = \nabla^*\nabla + (R/4)$$

where ∇ is the Levi-ivita connection, ∇^* its adjoint and R is the scalar curvature.

The surprising element here is the simplicity of the curvature term. Only the simplest invariant, namely the scalar curvature appears. As we will see in the proof below, this is partly a consequence of the fact that the spin representation is very "democratic" in the sense that all weights are equal.

The Lichnerowicz formula implies that a compact spin manifold with positive scalar curvature has no non-zero harmonic spinors. As a consequence, by the index theorem, compact spin manifolds with a non-zero $\hat{A}$-genus do not carry metrics of positive scalar curvature. Even some exotic spheres do not admit any metric with positive scalar curvature, according to Hitchin's [Hi] generalization of Lichnerowicz. The study of manifolds admitting a metric of positive curvature has since then become a huge area of research. Also it is interesting to note that the local curvature expressions for the characteristic class $\hat{A}(M)$ appearing as the local density of the heat kernel in the index theorem are quite elaborate and it is not easy to see why the simple condition $R > 0$ should imply directly that the top form in the index density is exact (on a compact manifold). Since, as explained above, the index density is the (super-)trace of the heat kernel $\exp(-tD^2)$ as $t \to 0^+$ whereas the harmonic projectors describe the behaviour of the heat kernel as $t \to \infty$, there should be a natural way to interpolate the index density (characteristic classes) and the Lichnerowicz formula by varying t in the heat kernel, so that the relatively complicated curvature expressions for the characteristic forms simplify (modulo exact forms) to a simple expression in the scalar curvature. One intriguing thing is that $\hat{A}$-genus depends only on the Pontryagin classes, whose local expressions in Chern–Weil theory depend only on the Weyl curvature and hence are "seemingly" independent of the scalar curvature, at least locally!

For the twisted Dirac operator with values in a vector bundle E, the Lichnerowicz formula for D^2 is computed to be:

$$D^2(\sigma \otimes \phi) = \nabla^*\nabla(\sigma \otimes \phi) + \frac{R}{4}\sigma \otimes \phi + \mathcal{R}(\sigma \otimes \phi) \tag{1}$$

for $\sigma \otimes \phi \in \Gamma(\mathbb{S} \otimes F)$, where $\nabla^*\nabla$ is the rough Laplacian, R is the scalar curvature, and the last term is explicitly given by:

$$\mathcal{R}(\sigma \otimes \phi) = \frac{1}{2}\sum_{j,k=1}^{m} \gamma(e_a)\sigma \otimes R^\nabla(e_a)\phi \tag{2}$$

where $\{e_a\}, a = 1, \ldots, m = \frac{n(n-1)}{2}$ is now an orthonormal base with respect to the metric g for the two-vectors $\bigwedge^2(T_pM)$ at the point in question, R^∇ is the curvature tensor of the connection in the bundle E, and γ is Clifford multiplication for g.

Proof. We define the second-order covariant derivative: $\nabla^2_{u,v} = \nabla_u\nabla_v - \nabla_{\nabla_u v}$ where for simplicity of notation, ∇ is used for all the covariant derivatives $\nabla^2_{u,v}$ is tensorial in u,v and its antisymmetric part is the curvature: $R(u, v) = \nabla^2_{u,v} - \nabla^2_{v,u}$ (here we use the fact that the Levi-Civita connection is torsion free). Now using a frame satisfying $\nabla_{e_k} e_l = 0$ at a given point we compute:

$$D^2 = \sum_{k,l} e_k \cdot \nabla_{e_k}\left(e_l \cdot \nabla_{e_l}\right)$$

$$= \sum_{k,l} e_k \cdot e_l \cdot \nabla_{e_k}\nabla_{e_l}$$

$$= \sum_{k=l} e_k \cdot e_l \nabla_{e_k}\nabla_{e_l} + 2\sum_{k<l} e_k \cdot e_l \nabla_{e_k}\nabla_{e_l}$$

$$= -\sum_k \nabla_{e_k}\nabla_{e_k} + \sum_{k<l} e_k \cdot e_l \cdot R(e_k, e_l).$$

The first term is the rough Laplacian $\nabla^*\nabla$ and the second term acting on $\sigma \otimes \phi$ can be simplified as:

$$\sum_{k<l} e_k \cdot e_l \cdot R(e_k, e_l)(\sigma \otimes \phi) = -\sum_{a=1}^{m} e_a \cdot \hat{R}(e_a)\sigma \otimes \phi + \sum_{a=1}^{m} e_a \cdot \sigma \otimes R^\nabla(e_a)\phi$$

where $\{e_a\}, a = 1, \ldots, m = \frac{n(n-1)}{2}$ is now an orthonormal base for $\bigwedge^2(TM)$, and $\hat{R}$ is the curvature operator of the Riemannian manifold. (Note the sign change!)

The second term is the same as $\mathcal{R}$ and the first term can be further simplified by choosing an orthonormal base $\{e_a\}$ for $\bigwedge^2(TM)$ that diagonalizes the curvature operator, so that $\hat{R}(e_a) = \lambda_a\, e_a$ and $\hat{R}(e_a)\sigma = \frac{1}{2}\lambda_a\, e_a \cdot \sigma$. It then follows that:

$$-\sum_{a=1}^{m} e_a \cdot \hat{R}(e_a)\sigma = -\frac{1}{2}\sum_{a=1}^{m} \lambda_a\, e_a \cdot e_a \cdot \sigma = \frac{R}{4}\sigma.$$

This proves the Lichnerowicz formula. □

For reference purposes, we would also like to record here a (generalized) Lichnerowicz formula for a metric connection with torsion:

$$D^2(\sigma \otimes \phi) = \left(\nabla^*\nabla + \frac{R}{4} + \mathcal{R}\right)(\sigma \otimes \phi) + \sum_{k,l} e_k \cdot e_l \nabla_{T(e_k, e_l)}(\sigma \otimes \phi) \quad (3)$$

where the extra term is a first-order operator involving the torsion T of the connection ∇. The proof is exactly the same as above, except for the fact that for a connection with torsion: $\nabla^2_{u,v} - \nabla^2_{v,u} = R(u,v) + \nabla_{T(u,v)}$.

2. Gromov's K-Area

2.1. Definition of K-Area

A general principle in Riemannian geometry states that large positive curvature should imply "small size". For Ricci curvature, this is made precise by the Bonnet–Myers theorem, which provides a sharp estimate for the diameter of a complete Riemannian manifold in terms of a positive lower bound for the Ricci curvature. This is of course no longer true (in dimensions larger than two) if the Ricci curvature is replaced by scalar curvature, since a Riemannian product of any Riemannian manifold with a sufficiently small S^2 has arbitrarily large positive scalar curvature. One expects therefore that a manifold with large positive scalar curvature is small in the sense that it is "close" to a codimension 2 subvariety.

Gromov's K-area measures a certain size of a Riemannian manifold that is related to scalar curvature in the sense that (at least for spin manifolds), large positive scalar curvature implies small K-area. The K-area is, roughly speaking, the inverse of the norm of the smallest curvature obtainable among all topologically essential unitary bundles equipped with connections on a given Riemannian manifold. K stands here both for K-theory and also for curvature (Krümmung). To measure the norm of the curvature, a metric g on the manifold is used. However, the definition does not involve the Riemannian curvature of the metric and hence the K-area is a pure C^0-invariant of g, or more precisely of the metric on 2-forms. It measures the K-theoretic two-dimensional size of the manifold. One can modify the definition by taking the supremum with respect to a suitable class of metrics, e.g., adapted metrics for a symplectic manifold, in order to get a symplectic invariant. One can also restrict to a special class of bundles and measure only certain parts of the curvature to get more refined invariants.

Since we would like to speak about symplectic connections, we will describe a rather general set-up to define the K-area. Let G be a connected Lie group, not necessarily finite-dimensional, whose tangent bundle is equipped with a bi-invariant norm defining a left invariant metric on G. We identify the Lie algebra $\mathfrak{g}$ of G with the space of right-invariant vector fields on G. Suppose that G acts on a connected manifold. The standard situation is when F is a vector space E and we have a representation of G. Another case which is also important is when F is a symplectic manifold and G is a (not necessarily finite-dimensional) subgroup of the infinite-dimensional group of all Hamiltonian symplectomorphisms: $Ham(F, \omega)$.

Consider a fiber bundle $\pi : P \to M$, over a Riemannian manifold (M, g) with fiber F associated to a G-principal bundle and G-connections on these bundles, i.e., connections whose parallel transports belong to the structural group G. Let R^∇ denote the curvature of a connection ∇ on the bundle P.

To a pair of vectors $v, w \in T_x M$, the curvature tensor associates an element $R^\nabla(v, w) \in \mathfrak{g}$, defining a G-vector field on the fiber $\pi^{-1}(x)$. The fiber $\pi^{-1}(x)$ can be identified with F, and since $\| \cdot \|$ is bi-invariant norm on $\mathfrak{g}$, $\| R^\nabla(v, w) \|$ is well defined independent of the identification. For $G = U(N)$ acting on a vector space E through a representation ρ, we will use the following supremum norm:

$$\|A\| = \max_{|u|=1} |\rho(A)(u)|,$$

where the maximum is taken over all unit vectors in E.

For a given bundle with a connection ∇ we define:

$$\|R^\nabla(F)\| = \sup_{|v \wedge w|=1} \|R^\nabla(v, w)\|,$$

where the maximum is taken over all unit bi-vectors $v \wedge w \in \Lambda^2(TM)$ with respect to the metric g.

Gromov's K-area of a compact even-dimensional Riemannian manifold (M^{2m}, g) is now defined by taking the supremum of $\|R^\nabla(E)\|^{-1}$ over all unitary bundles E with finite-dimensional structure groups $U(N)$ (all N), that have a non-vanishing Chern number and using linear connections ∇. That E has a non-vanishing Chern number is equivalent to the fact that the classifying map for E: $\chi_E : M \longrightarrow BU(N)$ is not homologous to zero. By an algebraic calculation involving Chern classes, it can be shown (see [G1]) that it is equivalent to the non-vanishing of the index of the Dirac operator twisted with some bundle E' that is associated to E, i.e., some tensor (or exterior) product of E with itself. We will call such bundles *homologically essential.*

Definition.
$$K\text{-area}\,(M^{2m}, g) = \sup_{E, \nabla} \|R^\nabla(E)\|^{-1},$$

where the maximum is taken over all homologically essential unitary bundles E of all dimensions and over all linear connections ∇.

Of course, instead of taking the infimum over all homologically essential bundles, one can also measure the K-area of individual bundles E, by taking the infimum over all unitary connections on E. This would define a norm on equivalence class of bundles, and hence a norm in K-theory.

Since the classifying space carries a universal connection on its universal bundle and every unitary connection is induced by a map into BU, one can think of minimizing the "surface area" among all homologically essential classifying maps.

In order to extend the definition to odd-dimensional manifolds we first define the K-area of a non-compact even-dimensional manifold exactly as above, except that we use bundles E which are trivial outside a compact set and also use characteristic classes with compact support. Now, for an odd-dimensional manifold (M^{2m+1}, g), we stabilize by taking products with $\mathbb{R}^{2k+1}$ for all k and define the stable K-area as:
$$K\text{-area}_{st}\,(M^{2m+1}, g) = \sup_k K\text{-area}\,(M \times \mathbb{R}^{2k+1}, g \times \bar{g}).$$

The K-area has some fundamental properties (see [G1] for more details):

(i) K-area scales like a two-dimensional area, and if $g_1 \geq g_2$ on 2-vectors, then $K\text{-area}\,(M, g_1) \geq K\text{-area}\,(M, g_2)$.

(ii) The K-area is strictly positive, since any even-dimensional manifold admits a unitary bundle with non-trivial top Chern class. This can be constructed locally in a small ball by pulling back the spinor bundle of S^{2n} with a degree one.

(iii) The K-area of a simply connected manifold is finite, since on a simply connected manifold an almost flat connection can be deformed to a flat connection. The monodromy of a flat connection along a closed loop of on a simply connected manifold is bounded by the norm of the curvature and an upper bound of the area of a null homotopy of the loop.

(iv) A finite covering (which is trivial outside a compact set) has the same K-area. This implies in particular that the K-area of a torus is ∞, since it can cover a multiple of itself by homotheties. More generally, the K-area of a closed manifold of non-positive sectional curvature whose fundamental group is residually finite is infinite.

That the torus has infinite K-area can also be seen from the fact that there are homologically essential bundles on a covering torus with arbitrarily small curvature. A large even-dimensional (covering) torus can always be mapped to a standard round sphere: $f : T^{2m} \to S^{2m}$ such that $|df|$ is very small. Now we can pull back the spinor bundle $\mathbb{S}^+$ on the sphere (this bundle has non-zero top Chern class and is the fundamental generator of the K-theory of even-dimensional spheres) to the torus via the map f. $f^*(\mathbb{S})$ is then homologically essential but will have arbitrarily small curvature on a sufficiently large torus, showing that the K-area of a torus is infinite.

2.2. The Fundamental Estimate in Terms of Scalar Curvature

Although the definition of K-area does not involve the curvature tensor of the Riemannian metric, there is a deep and perhaps surprising connection to the scalar curvature. In fact, part of the motivation that led Gromov to this new invariant is to give a new interpretation of the proof given by Gromov and Lawson of the following fundamental global theorem on scalar curvature rigidity of the torus.

Theorem 1. *Let g be a Riemannian metric on T^n with scalar curvature $R(g) \geq 0$ everywhere. Then g is flat.*

This was proved first by Schoen and Yau [SY1] for low dimensions (≤ 7) and then by Gromov and Lawson [GL1, GL2, GL3] in all dimensions. The proof by Schoen and Yau is a somewhat simpler version of their argument to establish the positive mass theorem and uses the second variation formula for stable minimal surfaces. The proof by Gromov and Lawson on the other hand, uses spinors and is closer in spirit to Witten's subsequent proof of the positive mass theorem. However, since the result is global, the argument is more elaborate than Witten's and is based on the index theorem. Gromov's definition of K-area gives an elegant re-interpretation of the basic idea of their proof, expressing it as a fundamental inequality relating the scalar curvature to the K-area.

The main technique is to obtain a bound for the K-area from above, in terms of the inverse of the infimum of the scalar curvature, provided the manifold carries a metric with positive scalar curvature. This can be achieved by analyzing the Lichnerowicz formula and using the index theorem. Since the torus has infinite K-area, we see that a torus cannot carry a metric of positive scalar curvature. Some extra geometric work is then needed to get the full scalar curvature rigidity result for the torus.

The fundamental K-area inequality of Gromov can now be stated as follows:

Theorem 2. *Every complete Riemannian spin manifold (M^n, g) with scalar curvature $R(g) \geq \kappa^2$ everywhere satisfies:*

$$K\text{-}area_{st}(M, g) \leq \frac{c(n)}{\kappa^2}$$

for some universal constant $c(n)$ depending only on the dimension.

The proof is an immediate consequence of the Lichnerowicz formula for twisted Dirac operators and the index theorem given the definitions. We refer again to [G1] for details.

Remark. For surfaces, the Gauss–Bonnet theorem (for S^1-bundles) implies a direct and exact relationship between the Euler characteristic and the Gaussian (= sectional) curvature. For example, in case of the round S^2, the K-area inequality is sharp for the Hopf bundle, which is the square root of the tangent bundle with Euler characteristic ± 1. It is a natural question now to find Riemannian manifolds which are "extremal" with respect to this K-area inequality. The first sharp result of this nature was obtained by Llarull [Ll1] who made a careful analysis of the proof by Gromov and Lawson in the case of the sphere to obtain the following theorem on the scalar curvature rigidity of spheres.

Theorem 3. *Let g be a Riemannian metric on S^n satisfying $g \geq \bar{g}$ on all 2-vectors and with scalar curvature $R(g) \geq R(\bar{g}) \equiv n(n-1)$ everywhere, where $\bar{g}$ is the standard metric of constant sectional curvature $K \equiv 1$. Then $g \equiv \bar{g}$ everywhere.*

This is the main result in Llarull's work [Ll1] In fact, Llarull proved a more general theorem [Ll2]. See also [Li2, Li3], where one can find various extensions and generalizations.

Theorem 4. *Let (M, g) be a complete Riemannian orientable spin manifold of dimension $n + 4k$ such that the scalar curvature satisfies $Sc(X) \geq n(n-1)$ and let $f : M \to S^n$ be a smooth non-strictly area decreasing map, i.e., satisfying $area(f(S)) \leq area(f(S))$ for all smooth two-dimensional surfaces $S \subset M$.*
*   If the pre-image $N_x = f^{-1}(x) \subset M$ of a generic point $x \in S^n$ satisfies $\hat{A}(N_x) \neq 0$ then $R(g) = n(n-1)$ and the map f is an isometric submersion.*

Theorem 3 basically says that round spheres are extremal for the K-area inequality among spin manifolds. We will provide below a simplified proof of Llarull's theorem in even dimensions. An appropriate modification of the proof yields the following results for compact symmetric spaces.

Theorem 5. *Let $(M^{2n}, \bar{g})$ be a compact Hermitian-symmetric space of constant scalar curvature $R(\bar{g})$ with Kähler form ω. If g is any Riemannian metric on M satisfying $|\omega|_g < |\omega|_{\bar{g}}$ then there is a point on M where the scalar curvatures satisfy $R(g) < R(\bar{g})$.*

Theorem 6. *Let $(M^{2n}, \bar{g})$ be an even-dimensional compact symmetric space of constant scalar curvature $R(\bar{g})$. Assume that either the Euler class or the signature* *of M is non-zero. If g is any Riemannian metric on M satisfying $g > \bar{g}$ on all 2-vectors, then there is a point on M where the scalar curvatures satisfy $R(g) < R(\bar{g})$.*

Remark. In a paper by S. Goette and U. Semmelmann [GS1], Theorem 4 has been generalized to Kähler manifolds with positive Ricci curvature. They show that these manifolds are extremal for the K-area inequality in the category of *spin^c*-manifolds. See also [Li3].

The assumption on the metrics in Llarull's theorem can be stated more geometrically by saying that all surfaces in M have larger area with respect to g than the standard metric $\bar{g}$. For Hermitian-symmetric spaces, we relax the assumption on the metrics and compare them only on the Kähler form, i.e., only the areas of holomorphic curves need to be compared.

The key step in obtaining sharp estimates for the K-area inequality is to find the optimal homologically essential twisting bundle for the Dirac operator. We are looking for an essential bundle with least curvature. We then have to make careful estimates of the curvature terms that appear in the Lichnerowicz formula and then appeal to the index theorem. In order to illustrate this we now present a rather detailed proof of Llarull's theorem for even-dimensional spheres. This is a somewhat simplified version of his proof and generalizes easily to prove Theorems 4, 5, 6 and other similar results.

For the purpose of local calculations, we may always assume, that the manifold is spin. Let $\mathbb{S}(g) = \mathbb{S}^+(g) \oplus \mathbb{S}^-(g)$ denote the bundle of spinors of an even-dimensional spin manifold (M^{2n}, g), so we have two spinor bundles with respect to the two metrics g and $\bar{g}$, where $\bar{g}$ is the standard metric. We consider the twisted Dirac operator D on the bundle $\mathbb{S}(g) \otimes E$, where we choose the coefficient bundle to be $E = \mathbb{S}^+(\bar{g})$ (or $\mathbb{S}^-(\bar{g})$), the spinor bundle with respect to the spherical, or more generally, a symmetric background metric $\bar{g}$. Here we use the metric g and its Levi-Civita connection to define the Dirac operator on the spinors in $\mathbb{S}(g)$, but for the twisting bundles $\mathbb{S}^{\pm}(\bar{g})$, the Levi-Civita connection of the metric $\bar{g}$ is used. We will regard the two spinor bundles $\mathbb{S}(g)$ and $\mathbb{S}(\bar{g})$ as isomorphic complex vector bundles over M, with two different metrics, but more importantly, admitting two different Clifford multiplications by vectors and exterior forms on M. To distinguish the two distinct Clifford multiplications, we will denote them by: $\sigma \mapsto \bar{\gamma}(v)\sigma$ for the metric $\bar{g}$ and $\sigma \mapsto \gamma(v)\sigma$ for the metric g, where v is a tangent vector (or more generally for $v \in \Lambda^*(TM)$).

After diagonalizing the metric g with respect to $\bar{g}$, so that we have two orthonormal bases: $\{\bar{e}_i\}$ for $\bar{g}$ and $\{e_i = \frac{1}{\lambda_i}\bar{e}_i\}$ for g, we can define $\gamma(e_i) = \bar{\gamma}(\bar{e}_i)$ and extend it canonically to the whole Clifford algebra with respect to g to give us a new representation on the same Hermitian vector space $\mathbb{S}$. γ then satisfies $\gamma(u)\gamma(v) + \gamma(v)\gamma(u) = -2g(u, v)$, and $\gamma(u)$ is skew adjoint.

The twisted Dirac operator is then given by:

$$D(\sigma_1 \otimes \sigma_2) = \sum_{k=1}^{2n} \left\{ \gamma(e_k)\nabla_{e_k}\sigma_1 \otimes \sigma_2 + \gamma(e_k)\sigma_1 \otimes \overline{\nabla}_{e_k}\sigma_2 \right\}$$

where $\{e_k\}$ is an orthonormal base for the tangent vectors with respect to the metric g, ∇ is the Levi-Civita connection of g, $\overline{\nabla}$ is the Levi-Civita connection of the metric $\bar{g}$, and $\sigma_1 \otimes \sigma_2 \in \Gamma(\mathbb{S} \otimes \mathbb{S}^+)$.

To simplify notation, we denote the product connection by ∇, i.e.:

$$\nabla_v(\sigma_1 \otimes \sigma_2) = \nabla_v\sigma_1 \otimes \sigma_2 + \sigma_1 \otimes \overline{\nabla}_v\sigma_2.$$

The last term in the Lichnerowicz formula:

$$D^2(\sigma_1 \otimes \sigma_2) = \nabla^*\nabla(\sigma_1 \otimes \sigma_2) + \frac{R}{4}\sigma_1 \otimes \sigma_2 + \mathcal{R}(\sigma_1 \otimes \sigma_2)$$

can be expressed as:

$$\mathcal{R}(\sigma_1 \otimes \sigma_2) = -\frac{1}{2}\sum_{a=1}^{m}\gamma(e_a)\sigma_1 \otimes \bar{\gamma}\left(\bar{R}(e_a)\right)\sigma_2$$

where $\{\bar{e}_a\}$, $a = 1,\ldots,m = n(2n-1)$ is an orthonormal base with respect to g for $\Lambda^2(T_pM)$, and $\bar{R}$ is the curvature operator of the symmetric metric. We note that the right-hand side is independent of the orthonormal base chosen and $\mathcal{R}$ is a well-defined self-adjoint algebraic operator on $\mathbb{S} \otimes \mathbb{S}$.

Let g now satisfy the condition: $g \geq \bar{g}$ on all 2-forms. This simply means that $g(v,v) \geq \bar{g}(v,v)$ for all $v \in \Lambda^2(TM)$. This implies that operator $\mathcal{R}$ dominates the corresponding operator $\overline{\mathcal{R}}$ defined by:

$$\overline{\mathcal{R}}(\sigma_1 \otimes \sigma_2) = -\frac{1}{2}\sum_{a=1}^{m}\bar{\gamma}(\bar{e}_a)\sigma_1 \otimes \bar{\gamma}\left(\bar{R}(\bar{e}_a)\right)\sigma_2$$

where $\{\bar{e}_a\}$, $a = 1,\ldots,m$ is now an orthonormal base for $\bar{g}$, in the sense that $\mathcal{R} - \overline{\mathcal{R}}$ is positive semi-definite on $\mathbb{S} \otimes \mathbb{S}$ with respect to the metric $g \otimes \bar{g}$ (and hence also w.r.t. $\bar{g} \otimes \bar{g}$). This is easily seen by choosing $\{e_a\}$ to be an orthonormal base of eigenforms that diagonalizes the metric g with respect to the background metric $\bar{g}$ on $\Lambda^2(T_pM)$, so that $e_a = \frac{1}{\lambda_a}\bar{e}_a$ with $\lambda_a \geq 1$ and with $\gamma(e_a) = \bar{\gamma}(\bar{e}_a)$. $\overline{\mathcal{R}}$ is then given by:

$$\overline{\mathcal{R}}(\sigma_1 \otimes \sigma_2) = -\frac{1}{2}\sum_{a=1}^{m}\frac{1}{\lambda_a}\left(\gamma(e_a)\sigma_1 \otimes \bar{\gamma}\left(\bar{R}(\bar{e}_a)\right)\sigma_2\right).$$

If $\mathcal{K}_p \subset \mathcal{O}(T_pM) = \Lambda^2(T_pM)$ denotes the holonomy (= isotropy) subalgebra of the symmetric space at the point in question, then the curvature operator of $\bar{g}$ is just the orthogonal projection onto $\mathcal{K}_p$ followed by an invertible symmetric map $S : \mathcal{K} \to \mathcal{K}$. Therefore:

$$\overline{\mathcal{R}}(\sigma_1 \otimes \sigma_2) = -\frac{1}{2}\sum_{a=1}^{m}\bar{\gamma}(\bar{e}_a)\sigma_1 \otimes \rho_a\bar{\gamma}(e_a)\sigma_2$$

where the ρ_a's are the non-zero eigenvalues of the curvature operator and $m = \dim(\mathcal{K})$. On each simple factor $\mathcal{K}_s$ of $\mathcal{K}$, the eigenvalues of the operator $\bar{R}$ are all equal $(= \rho_s \neq 0)$, so

$$\overline{\mathcal{R}}_s(\sigma_1 \otimes \sigma_2) = -\frac{\rho_s}{2} \sum_{a=1}^{m_s} \bar{\gamma}(\bar{e}_a)\sigma_1 \otimes \bar{\gamma}(\bar{e}_a)\sigma_2$$

where $m_s = \dim(\mathcal{K}_s)$, and $\{\bar{e}_a\}$ is an orthonormal base for $\mathcal{K}_s \subset \Lambda^2(T_pM)$. $\overline{\mathcal{R}}_s$ is independent of the choice of the orthonormal base.

We now estimate the minimum eigenvalue of the operator $\overline{\mathcal{R}}_s$. The Casimir operator $\mathcal{C}_s$ of the representation on $\mathbb{S} \otimes \mathbb{S}$ induced by the isotropy representation restricted to each simple component $\mathcal{K}_s$ is given by:

$$\mathcal{C}_s(\sigma_1 \otimes \sigma_2) = -\sum_{a=1}^{m_s} (\bar{\gamma}(\bar{e}_a))^2 (\sigma_1 \otimes \sigma_2)$$

$$= -\sum_{a=1}^{m_s} \left\{ (\bar{\gamma}(\bar{e}_a))^2 \sigma_1 \otimes \sigma_2 + 2\bar{\gamma}(\bar{e}_a)\sigma_1 \otimes \bar{\gamma}(\bar{e}_a)\sigma_2 + \sigma_1 \otimes (\bar{\gamma}(\bar{e}_a))^2 \sigma_2 \right\}$$

$$= 2m_s(\sigma_1 \otimes \sigma_2) + \frac{4}{\rho_s}\overline{\mathcal{R}}_s(\sigma_1 \otimes \sigma_2)$$

where we used the Clifford identity $(\bar{\gamma}(\bar{e}_a))^2 = -Id$. Now $\mathcal{C}_s$ is positive semi-definite on $\mathbb{S} \otimes \mathbb{S} \approx \Lambda^*(TM)$. In fact, it is positive definite on all non-trivial irreducible components of the representation and is equal to zero only on the trivial representations that occur, in particular for Λ^0 and Λ^n. In any case, we have the following basic algebraic estimate:

$$\overline{\mathcal{R}}_s \geq -m_s \frac{\rho_s}{2} Id.$$

This implies:

$$\mathcal{R} \geq \overline{\mathcal{R}} = \sum_s \overline{\mathcal{R}}_s \geq -\sum_s m_s \frac{\rho_s}{2} Id = -\frac{\overline{R}}{4} Id \geq -\frac{R}{4} Id$$

provided $R = R(g) \geq R(\bar{g}) = \overline{R}$. Moreover, $\mathcal{R}$ is strictly $> -\frac{R}{4} Id$ unless all the inequalities above are strict equalities everywhere on M. This yields, by the Lichnerowicz formula, a vanishing theorem for both $\mathbb{S}^+(\bar{g})$ and $\mathbb{S}^-(\bar{g})$-valued harmonic spinors on M, provided that $g \geq \bar{g}$ on $\mathcal{K}_p \subset \Lambda^2(T_pM)$ for each p, and $R(g) \geq R(\bar{g})$ everywhere, with strict inequality holding at least at one point.

By the index theorem the topological indices of the two twisted Dirac operators are given by:

$$\int \hat{A}(M) \cdot ch(\mathbb{S}^\pm) = \frac{1}{2}(e(M) \pm \tau(M)) \tag{4}$$

where $e(M)$ is the Euler characteristic and $\tau(M)$ is the signature of M.

This proves Theorems 3 and 6, since the twisted bundles are globally defined for any Riemannian manifold, and we need the factors (the spinor bundles) only for local calculations. In the case of the sphere, the curvature operator is just the identity map on 2-vectors, so rigidity follows from the fact that if strict equality holds in the inequalities above we must have $g \equiv \bar{g}$ on $\Lambda^2{}_p$ at all points. Since $g \geq \bar{g}$ on Λ^2 this now implies that the two metrics are identical on each tangent space. The odd-dimensional spherical case can be proved by applying the even-dimensional case to $M \times S^1$, which is given an appropriate metric with a long S^1. We refer to [Ll1]. A slightly more careful analysis of the curvature terms in our proof will show that the proofs can be extended to the case where the curvature operator has positive eigenvalues.

To prove Theorem 5, we consider the twisted Dirac operator D on the bundle $\mathbb{S}(g) \otimes E$, where we choose the coefficient bundle E to be the line bundle $L(\bar{g})$, whose square is the canonical bundle of the Hermitian symmetric space. Here we use the metric g and its Levi-Civita connection to define the Dirac operator on the spinors in $\mathbb{S}(g)$, but for the twisting bundle $L(\bar{g})$, the connection induced by the Levi-Civita connection of the symmetric metric $\bar{g}$ is used. Under the assumption that $|\omega|_{\bar{g}} > |\omega|_g$ the operator $\mathcal{R}$ dominates the corresponding operator $\overline{\mathcal{R}}$:

$$\mathcal{R}(\sigma \otimes l) = -\frac{\overline{R}}{4|\omega|_g^2}\bar{\gamma}(\omega)\sigma \otimes \bar{\gamma}(\omega)l$$

which in turn dominates $-\overline{R}/4$ so that $\mathcal{R} + \overline{R}/4$ is positive definite on $\mathbb{S} \otimes L$. The assumption $R \geq \bar{R}$ would then imply that there are no harmonic spinors by the Lichnerowicz' formula. The tensor product of the spinor bundle with the line bundle whose square is the dual of the canonical bundle exists globally on any Hermitian manifold, and is the spinor bundle of the canonically associated *spin^c* structure. The spinor bundle might not be globally defined and the square root of the canonical bundle might not globally exist but their tensor product is well defined, because the sign ambiguities cancel exactly! The index for the corresponding Dirac operator is given by the Todd genus which is non-zero for compact Hermitian symmetric spaces, and Theorems 5 and 6 follow.

Our proofs above can be easily extended in various directions as has been done in [GS2, Li2, Li3]. For example, the following can be proved.

Theorem 7. *Let (M^{2m}, g) be a compact spin manifold and let $f : (M, g) \rightarrow (\mathbb{C}P^m, \hat{g})$ be a map, where $\hat{g}$ is the standard metric. Assume m is even (so that $\mathbb{C}P^m$ is spin). If the scalar curvature of (M, g) at any point $x \in M$ is not smaller than the scalar curvature of $(\mathbb{C}P^m, \hat{g})$ at the corresponding mage point $f(x) \in \mathbb{C}P^m$, the f is an isometry.*

It should also be pointed out that scalar curvature is extremely "flexible" in the opposite direction (upper bounds), and changing the sign of the inequality in the above theorems has completely different consequences upper bounds for scalar curvature, or even Ricci curvature exhibit even "local flexibility" instead of "global rigidity" for lower bounds as was shown first by J. Lohkamp [Lo1, Lo2]. There even exist local deformations of the metric which decrease the scalar curvature.

Theorem 8. *Let (M, g) be a complete Riemannian manifold of dimension $n \geq 3$, and let ψ be a smooth function on M such that $\psi(x) \leq R_g(x)$ for each point*

$x \in M$. Let $U = \{x \in M : \psi(x) < R_g(x)\}$, and let U_ϵ denote an ϵ-neighbourhood of the set U. Given any $\epsilon > 0$, there exists a smooth metric $\hat{g}$ on M such that $\psi(x) - \epsilon \leq R_{\hat{g}}(x) \leq \psi(x)$ at each point in U_ϵ and $\hat{g} = g$ outside U_ϵ.

Even for the Ricci curvature we have the following rather amazing result:

Theorem 9. *Any Riemannian metric g on a manifold of dimension ≥ 3 can be C^0-approximated by a metric with scalar curvature satisfying $\mathrm{Ric} \leq -(n-1)$.*

2.3. Connections with Symplectic Invariants

In symplectic geometry, there is a notion of fibrations $\pi : P \to M$ with a symplectic manifold F as fiber, where the structure group is the group of (exact) Hamiltonian symplectomorphisms of the fiber. These are called symplectic fibrations. If the base manifold (M, ω_M) is also symplectic, there is a weak coupling construction, originally due to Thurston, of defining a symplectic structure on the total space P. An efficient way to describe this procedure is through the use of the curvature of a symplectic connection and results in what is known as minimal coupling form. We refer to [GLS] for details. Parallel translation w.r.t. a symplectic connection is a symplectomorphism of the fiber, and hence the symplectic curvature can be described as a two-form on the base manifold M with values in the Hamiltonian vector fields on the fiber at the given point and hence can be identified with a function (the Hamiltonian) on the fiber. We normalize Hamiltonians on compact symplectic manifolds to have mean value zero. If the fiber F is compact and simply-connected, each symplectic connection Γ gives rise to a unique *closed* 2-form ω^Γ on the total space P characterized by the following properties:

(i) ω^Γ restricts to the symplectic form on the fibers.

(ii) The horizontal space and the vertical space of the connection Γ are perpendicular w.r.t. ω^Γ.

(iii) On the horizontal space, ω^Γ coincides with the symplectic curvature of the connection Γ.

(iv) $\pi_*(\omega^\Gamma)^{d+1} = 0$, where π_* is integration over the fiber (Gysin map) and $2d$ is the dimension of F

The cohomology class of ω^Γ is independent of the connection Γ and hence is a symplectic invariant of the fibration. In general, the form ω^Γ is not symplectic (it could be degenerate in horizontal flat directions). However if we define the weak coupling form:

$$\omega_\epsilon = \epsilon \omega^\Gamma + \pi^*(\omega_B),$$

then for sufficiently small ϵ, this would define a closed non-degenerate symplectic form on the total space P. The maximal possible value ϵ_{max} is a symplectic invariant called the maximal weak coupling constant.

Using this interpretation of symplectic curvature, we can now define the symplectic K-area of a given *fixed* symplectic fibration over a compact symplectic as the inverse of the minimum possible curvature. First we define the norm:

$$\|\omega^\Gamma\| = \max_{v,w} \frac{|\omega^\Gamma(\bar{v}, \bar{w})|}{|\omega_B(v, w)|}$$

where $\bar{v}, \bar{w}$ denote the horizontal lifts and the maximum is taken over all pairs of vectors v, w in the base manifold B such that $\omega_B(v, w) \neq 0$.

The symplectic K-area is then defined to be:

Definition.

$$K_{\text{symp}}\text{-area}\,(P) = \sup_\Gamma \|\omega^\Gamma\|^{-1}$$

where the maximum is taken over all symplectic connections on the fixed fibration: $P \to M$.

In many situations we are dealing with symplectic fibrations that arise from linear vector bundles. An important case is when the fiber is a co-adjoint orbit $F = \mathcal{O} \subset \mathfrak{g}^*$ and the bundle is associated to a principle G-bundle via the adjoint representation. G is here a finite-dimensional Lie group, e.g., $U(n)$. An "ordinary" connection ∇ is then necessarily symplectic since G acts symplectically on the co-adjoint orbits with moment map given by the inclusion: $F \subset \mathfrak{g}^*$. The symplectic curvature of ∇ is therefore simply the curvature $R^\nabla \in \mathfrak{g}$ thought of as a linear function on $\mathfrak{g}^*$ restricted to the co-adjoint orbit. It is clear that the symplectic K-area is an upper bound for the "ordinary" K-area (of the fixed bundle), since we are taking the smallest possible curvature among a larger class of connections. Optimistically, one might expect that for simple co-adjoint orbits, the two K-areas are equal, since the curvature of "ordinary" connection induces a simple linear Hamiltonian function of the fiber and so should be minimizing among all symplectic connections. That this is in fact true, was shown by L. Polterovich [Pol] for the special case of certain complex projective bundles over S^2. Moreover, it is not hard to see that the symplectic K-area is bounded from above by the maximal ϵ_{max} for the weak coupling constant, and Polterovich was able to show that all three invariants are equal in these special cases. This allows him also to conclude a sharp estimate for the Hofer norm of some loops in $U(n+1)$ acting on $\mathbb{C}P^n$. His proof uses the theory of J-holomorphic curves and Gromov–Witten invariants on the total space of the fibration. It would be intriguing to see whether there is a simple "spinorial" proof in the spirit of the last section of these results and also more generally investigate when symplectic K-area inequalities are sharp.

2.4. The Vafa–Witten Inequality

In 1984 Vafa and Witten [VW] proved the following surprising fact about the spectrum of twisted Dirac operators on compact spin manifolds. These results do not hold for ordinary Laplacians on bundle-valued forms. Simple counterexamples can be found by looking at line bundles on Riemann surfaces of sufficiently high degree. The inequalities that are proved also go in the opposite direction to the usual problem of finding lower bounds for the first eigenvalue that is usually considered in Riemannian geometry, for example, by employing the Bochner technique. In general. it seems to be harder to find lower bounds than upper bounds, say for the first eigenvalue. Upper bounds can be more easily established by an appropriate "test function"! The novelty in this case is the fact that the bounds here are independent of the twisting bundle (external symmetry group) and depends only on the internal metric of the manifold. Vafa and Witten were motivated by physical ideas about the absence of mass gaps and paramagnetic inequalities for fermions in quantum chromodynamics but their main result can be stated as follows:

Theorem 10. *Let* $|\lambda_1| \leq |\lambda_2| \leq \cdots$ *be the eigenvalues ordered by their absolute values of a twisted Dirac operator* D_E *defined on* $\mathbb{S} \otimes E$ *over a compact spin manifold* $(M.g)$ *of dimension* n, *where we use a connection* ∇ *for* E. *Then there exists a constant* $C(M,g)$, *depending only on the Riemannian manifold* $(M.g)$ *and independent of the twisting bundle* E *and the connection* ∇ *on* E *such that for all* k *we have the following universal bound:*

$$\frac{1}{k}|\lambda_k|^n \leq C(M,g)$$

For odd-dimensional manifolds, there are stronger results (which do not hold in even dimensions) to the effect that every interval on the real line of a certain length C (depending only on (M,g) but not on the twisting bundle E nor on the connection ∇ used on E) contains an eigenvalue of D_E.

Proof. For the sake of clarity, we will restrict ourselves to the case $k = 1$, i.e., the first eigenvalue and to even-dimensional manifolds first. If index$(D_E^+) \neq 0$, then there is nothing to prove since there is a harmonic spinor (zero eigenvalue) for D_E. The strategy is to show that any D_E is close (up to an algebraic zeroth-order operator) to a twisted Dirac operator which has non-zero index. To be more precise we show that this is true for the twisted Dirac operator of some multiple $E \otimes \mathbb{C}^N$ of E which has the same spectrum as D_E up to multiplicities. This is done in two steps:

Step 1. Find a bundle F such that index$(D_{E \otimes F}^+) \neq 0$

Step 2. Find a complementary bundle $F^\perp$ such that $F \oplus F^\perp$ is trivial.

Step one is achieved in the usual fashion, by pulling back to M the spinor bundle $\mathbb{S}^+$ of S^{2m}, whose top Chern class is non-zero, using a map of degree one: $M^{2m} \to S^{2m}$. (We can just map a small ball onto the sphere punctured at a point and the rest of the manifold can be mapped to that point.) This makes the index of $D_{E \otimes F}^+$ equal to $\dim(F)$, since we are assuming that index$(D_E^+) = 0$.

For step two we can choose the complementary bundle to be the pull back (under the same map as in step one) of the universal complementary bundle for $\mathbb{S}^+$ on the sphere (this is simply the bundle $\mathbb{S}^-$ on the sphere). Moreover we can use the connections that are also pull backs of the standard spinor connections on the round sphere. The difference of the connection on $E \otimes (F \oplus F^\perp) \cong E \otimes \mathbb{C}^N$ defined by these pull-backs to that induced by the original connection ∇ on the multiple $E \otimes \mathbb{C}^N$ now depends only on the universal bundles on the sphere and the map f (in fact, only on $|\wedge^2 df|$) that is used to pull back the bundles and hence is independent of E and ∇. A suitable multipole of the original Dirac operator is therefore at a bounded distance from another twisted Dirac operator with non zero index. (The difference between the two operators is a zeroth-order term controlled by the difference of the connections involved.) As a consequence, the spectrum of the original Dirac operator D_E is close, i.e., at a bounded distance, from an operator which has a zero eigenvalue, and the theorem is proved in this case.

To extend the theorem to higher eigenvalues (higher k) we need to use higher degree maps by applying the above construction to a union of disjoint balls, i.e., we pull back the bundles from the sphere and graft them onto the manifold on many small disjoint balls. This will give us get bigger indices (= degree $\times \dim(F)$) for the twisted Dirac operators and hence harmonic spinors of high multiplicity allowing us to control higher eigenvalues of twisted Dirac operators that are at a bounded distance from one with a big kernel.

To prove the odd-dimensional case, we take the product of the odd-dimensional manifold with S^1 to reduce it to the even-dimensional case, but to obtain the sharper result on spectral gaps is based on a spectral flow argument. This proves the more general result about the distribution of eigenvalues in an interval (not necessarily containing zero) of a definite length. $\square$

Although the proof seems rough and topological, the Vafa–Witten upper bounds are quite sharp in certain cases. Vafa and Witten themselves determined sharp bounds for flat tori. For spheres, and presumably other symmetric spaces, the estimates are sharp and the method is in fact closely related to the proof of Llarull's theorem. To be more precise, we can prove the following result:

Theorem 11. *Let $|\lambda_1| \leq |\lambda_2| \leq \cdots$ be the eigenvalues ordered by their absolute values of a twisted Dirac operator D_E defined on $\mathbb{S} \otimes E$ over the sphere S^n with the standard metric of constant sectional curvature 1, where we use a connection ∇ for E. Then*

$$k^{-\frac{1}{n}}|\lambda_k| \leq \frac{n}{2}$$

for all twisting bundles E and connections ∇ on E.

The inequality is sharp and the upper bound is attained for the trivial twisting bundle (i.e., for ordinary spinors). To get sharp estimates for $M = \mathbb{C}P^n$ (n odd) one should perhaps use the trivial bundle $(TM \oplus 1_{\mathbb{C}}) \otimes H$ where H is the Hopf bundle and $1_{\mathbb{C}}$ is the trivial complex line bundle.

Proof. If either index$(D_E^+) \neq 0$ or index$(D_E^-) \neq 0$ then $\lambda_1 = 0$. If this is not the case, we twist E with one of the spinor bundles $\mathbb{S}^\pm$ in even dimensions (and with the spin bundle $\mathbb{S}$ in odd dimensions). We use the connection that comes from the standard metric on the sphere for these spinor bundles as in the proof of Llarull's theorem of the last section. The complementary bundle that we use to trivialize the connection is $\mathbb{S}^\mp$ in the even-dimensional case and $\mathbb{S}$ in odd dimensions. The natural trivialization and flat connection of the sum $\mathbb{S}^\pm \oplus \mathbb{S}^\mp$ (resp. $\mathbb{S} \oplus \mathbb{S}$) comes from the fact that the sphere is isometrically imbedded in flat Euclidian space and these bundles are restrictions of the spinor bundles of $\mathbb{R}^{n+1}$ to the sphere. We can then use the Lichnerowicz formula as in the last section to show that $|\lambda_1| \leq \frac{n}{2}$. To obtain estimates for higher eigenvalues we twist with appropriate powers of the spinor bundles in order to increase the index. □

The main idea used behind the Vafa–Witten proof can be formalized to the notion of *K-length* of a Riemannian manifold. For a given bundle E with connection, we can measure its non-triviality by finding the minimum amount of "second fundamental form" required to imbed it inside a larger trivial bundle with a flat connection. To be more precise, we define:

$$\alpha(E, \nabla) = \inf \|A(\nabla, \bar{\nabla})\|$$

where the infimum is taken over all bundles $\bar{E}$ with a flat connection $\bar{\nabla}$ which contains E as a subbundle, and $\|A\|$ is the norm of the second fundamental form which measures the difference between the connection ∇ and the connection induced from $\bar{\nabla}$ by projection onto the subbundle. Note that every vector bundle is a subbundle of some higher-dimensional trivial bundle!

The K-length of an even-dimensional Riemannian manifold (M, g) is then defined to be:

$$K\text{-length}(M, g) = \sup\left(\alpha(E, \nabla)\right)^{-1}$$

where the supremum is taken over all homologically essential unitary bundles E with connection ∇.

The extension of this definition to odd-dimensional manifolds proceeds exactly as in the case of the K-area. Large K-length implies large K-area, but the converse is not necessarily true. The Vafa–Witten technique explained above can be extended to other situations where the K-length can be used to control spectral gaps (this was the original reason why Vafa and Witten established their result). We refer again to [G1] for more details. For some other results related to this section, we refer to [D, DM, Go]. I suggest that one should make a more systematic investigation of the spin geometry of all compact symmetric spaces, even the ones that are not spin (by twisting with an appropriate bundle).

3. Positive Mass Theorems

3.1. Description of Results

Many of the classical concepts of Newtonian physics, such as mass, energy and momentum are ill defined in Einstein's theory of general relativity. There is no satisfactory notion of total energy, since the energy of the gravitational field itself is described purely in terms of geometry and does not contribute directly to the local stress-energy-momentum tensor T_{ij}. The problem here is that the geometry of space-time itself is a dynamic variable and a measurement is in principle always "relative" and hence one has to "take a stand and break the symmetry" in order to have a reasonable notion of mass and energy. The most natural situation occurs when gravitational forces are weak almost everywhere, except for a confined isolated region. More precisely, in an asymptotically flat space time describing an isolated system like a star or a black hole, where the gravitational field approaches ordinary Newtonian gravity with respect to a background inertial coordinate system at infinity, one can define the total mass, or more relativistically, the total energy-momentum four-vector of the system by asymptotic comparison with Newtonian theory at large distances. One also expects this four-vector to be "asymptotically" a conserved quantity as in Newtonian gravity. Although Hermann Weyl was the first person to propose a tentative definition of the energy and mass of an isolated system in his book: "Space, Time and Matter", it was Arnowitt, Deser and Misner who later gave a more precise definition.

More precisely, we define an asymptotically Euclidean space-like hypersurface to be a three-dimensional oriented Riemannian manifold (M, g), isometrically imbedded in four-dimensional Lorentzian space-time, whose first and second fundamental forms g_{ij} and h_{ij} satisfy the following asymptotic conditions:

(A) There is a compact set $K \subset M$ so that $M \setminus K$ is a finite disjoint union of ends, each diffeomorphic to the complement of a closed ball in $\mathbb{R}^3$, and using the standard coordinates given by this diffeomorphism, g and h have the asymptotic behaviour:

$$\partial_\alpha(g_{ij} - \delta_{ij}) \in O(r^{-1-|\alpha|}) \quad \text{for } |\alpha| \leq 2 \tag{5}$$

and

$$\partial_\beta(h_{ij}) \in O(r^{-2-|\beta|}) \quad \text{for } |\beta| \leq 1. \tag{6}$$

Here α, β are multi-indices. These are not the optimal decay rates, and we refer to [Ba] for refinements and also for a discussion of the independence of the ADM-mass, from the choice of the coordinate system at infinity. ADM stands for Arnowitt, Deser and Misner and their definition of the total energy-momentum (E, P_j) of an asymptotically Euclidean space-like slice is:

$$E = \frac{1}{16\pi G} \lim_{r \to \infty} \oint_{S(r)} (\partial_k g_{ik} - \partial_i g_{kk}) d\sigma^i \tag{7}$$

$$P_j = \frac{1}{8\pi G} \lim_{r \to \infty} \oint_{S(r)} (h_{ij} - \delta_{ij} h_{kk}) d\sigma^i \tag{8}$$

where $S(r)$ is the Euclidean sphere of radius r and the integrals are defined for each end. For the important prototypical example of the Schwarzschild metric:

$$ds^2 = \left(1 - \frac{2M}{r}\right) dt^2 + \left(1 - \frac{2M}{r}\right)^{-1} r^2 dr d\theta^2,$$

defined for $r > 2M$ this definition of course recovers the usual mass M that appears in the metric and $P = 0$.

The next important physical assumption assumption is the following dominant energy condition for the local mass density T:

(B) For each time-like vector e_0 transversal to M, $T(e_0, e_0) \geq 0$ and $T(e_0, .)$ is a non-space-like covector.

This implies that for any adapted orthonormal frame (e_0, e_1, e_2, e_3) with e_0 normal and e_1, e_2, e_3 tangential to M, we have the inequalities:

$$T^{00} \geq |T^{\mu\nu}| \text{ for all } 0 \leq \mu, \nu \leq 3 \tag{9}$$

and

$$T^{00} \geq (-T_{0k}T^{0k})^{1/2} - \tag{10}$$

We of course also assume that space time satisfies Einstein's field equations:

$$R_{\mu\nu} - \frac{1}{2}Rg_{\mu\nu} = 8\pi G T_{\mu\nu}. \tag{11}$$

Given these assumptions, the positive energy theorem states that

Theorem 12. *An asymptotically Euclidean space-like hypersurface in a space-time satisfying Einstein's equation, and the dominant energy condition has non-negative total energy in the sense that $E \geq |P|$ for each end. Moreover, if $E = 0$ for some end, then there is exactly one end, and M is isometric to the flat Euclidean space.*

After several attempts by relativists, who established several special cases, the first complete proof of this result was achieved by Schoen and Yau [SY2, SY3, SY4], using minimal surface techniques. Subsequently, Witten [W1] found a completely different proof using ideas from super-gravity involving harmonic spinors (some of the technical analytical details in Witten's paper were clarified later by Parker and Taubes [PT]).

In the positive mass theorem, if we ignore the fact that the space-like slice (M^3, g) is imbedded in space-time, the assumption to be asymptotically Euclidean is well defined by (1), and the formula (3) for the energy (or mass) E still makes sense. In this Riemannian situation, the appropriate assumption that replaces the dominant energy condition (B) is that the scalar curvature of (M^3, g) is non-negative. In fact, if the space-like slice has zero mean curvature zero $(tr(h) = 0)$, then by the Gauss–Codazzi equations, assumption (B) would imply that the scalar curvature of M^3 is non-negative. More generally, one would like to pose the problem whether, for all dimensions n, an asymptotically Euclidean Riemannian manifold (M^n, g) with non-negative scalar curvature has non-negative mass. This is also pertinent to physics, if one works in a more general framework than classical relativity. To be exact, one needs to modify the decay rate in definition (1), depending on n. We assume for simplicity:

$$\partial_\alpha(g_{ij} - \delta_{ij}) \in O(r^{-n+2-|\alpha|}) \quad \text{for } |\alpha| \leq 2. \tag{12}$$

With this definition of asymptotically Euclidean we have the following result.

Theorem 13. *An asymptotically Euclidean spin manifold M with non-negative scalar curvature everywhere has positive total mass E. Moreover $E = 0$ if and only if M is isometric to flat Euclidean space.*

Both Schoen–Yau and Witten established this theorem in dimension 3 in the course of their proof of the positive mass conjecture. For general n, this was first proved by Bartnik [Ba]. We note that for a compactly supported perturbation of the flat metric, the above theorem would be a simple consequence of the corresponding result for the torus.

It is a natural question to ask for similar results with different asymptotic background geometries and in [M1], I established, using Witten's method, the following hyperbolic version of the rigidity part of the last theorem.

Theorem 14. *A strongly asymptotically hyperbolic spin manifold of dimension > 2, whose scalar curvature satisfies $R \geq -n(n-1)$ everywhere, is isometric to hyperbolic space.*

There is a technical mistake in the definition of strongly asymptotically hyperbolic in my paper [M1], and I would like to correct it here.

Definition. A Riemannian manifold (M^n, g) is said to be *strongly asymptotically hyperbolic* (with one end) if there exists a compact subset $B \subset M$ and a diffeomorphism $\phi : M \setminus B \to H^n \setminus \bar{B}(r_0)$ for some $r_0 > 0$ such that, if we define the gauge transformation $A : T(M \setminus B) \to T(M \setminus B)$ by the equations:

(i) $g(Au, Av) = \phi^* g(u, v)$,   (ii) $g(Au, v) = g(u, Av)$,

then A satisfies the following properties:

AH1: There exists a uniform Lipschitz constant C such that

$$C^{-1} \leq \inf_{|v|=1} |Av| \leq \sup_{|v|=1} |Av| \leq C.$$

AH2: $\exp(\phi \circ r)(A - \mathrm{id}) \in L^{1,1} \cap L^{1,2}(T^* \otimes T(M \setminus B))$.

Before I sketch the main ideas behind the spinorial proofs of these theorems here are some remarks:

1. There is a much stronger version of the positive mass conjecture known as the Penrose conjecture for black holes where the mass is bounded from below by the area of the event horizon. The Riemannian version of this conjecture has been now established by Huisken–Ilmanen [HI] and H. Bray [Br]. The proofs are not spinorial and use what is known as the inverse mean curvature flow. For a spinorial proof of a related "Penrose-like inequality" see [He2].

2. There are also Lorentzian versions of these results for asymptotically Anti-de Sitter spaces. See for example: [RT, Wo].

3. There are some recent new positive energy conjectures formulated by Horowitz and Myers [HM], in their attempt to deal with stability problems arising from the AdS/CFT correspondence. For example, they conjecture a positive energy theorem and a corresponding rigidity result for metrics on four-dimensional manifolds which asymptotically look like:

$$ds^2 = \frac{r^2}{l^2}\left[\left(1 - \frac{r_0^4}{r^4}\right)d\theta^2 + (dx^1)^2 + (dx^2)^2\right] + \left(\frac{r^2}{l^2}\left(1 - \frac{r_0^4}{r^4}\right)\right)^{-1} dr^2$$

where $r \geq r_0$ and $\theta \in S^1$ with period $\pi l^2/r_0$. This the Euclidean version of their conjecture, and energy has to be appropriately defined.

4. Needless to say a lot of things have happened since I first published my result about asymptotically hyperbolic spaces in 1989, 30 years ago! I was not able to keep track of everything that was done since then, but a number of researchers have generalized my result in various directions. They have also introduced various notions of "hyperbolic mass at infinity". Here is a selection of some papers related to these topics [ACG, CH, Wa, Z].

5. Finally, in analogy with the flat case where the scalar curvature rigidity theorem for the torus can be seen as the "global" compact version of the positive mass theorem, which establishes the "local" scalar curvature rigidity of flat Euclidean space, one would like to have the following result for compact hyperbolic manifolds:

Conjecture. *Let g be a Riemannian metric on a compact quotient of hyperbolic space $M = \Gamma \backslash H^n$ of dimension $n \geq 3$ satisfying $vol(M, g) = vol(M, \bar{g})$ where $\bar{g}$ is the hyperbolic metric of constant sectional curvature -1. If the scalar curvature satisfies: $R(g) \geq R(\bar{g}) \equiv -n(n-1)$ everywhere, then g is isometric to $\bar{g}$.*

This is a special case of a more general conjecture of Gromov:

Conjecture. *Let M^n be a complete Riemannian manifold with sectional curvature $K \leq 1$ and let N^n be a compact Riemannian manifold with scalar curvature $R \geq -n(n-1)$. Then every continuous map $f_0 : M \to N$ is homotopic to a map f_1, such that $vol(f_1(M)) \leq vol(N)$. Moreover, this inequality is strict, unless N has a constant negative curvature and the map f_0 is homotopic to a locally isometric map.*

This would be a vast generalization of the Mostow rigidity theorem for hyperbolic metrics and also of the beautiful result involving entropy and negative Ricci curvature by P. Besson, G. Courtois and S. Gallot [BCG].

3.2. Main Ideas behind the Proofs

The first basic step in proving these theorems is to solve for harmonic spinors which have the correct behaviour at infinity and then apply the integrated version of the Lichnerowicz formula. Using now Stokes' theorem, we obtain boundary integrals, which in the asymptotic limit, are then identified with the "mass".

If we apply Stokes' theorem on a manifold with boundary to the Lichnerowicz formula for the ordinary Dirac operator:

$$D^2 = \nabla^*\nabla + \frac{R}{4}$$

we obtain:

$$\int_M \left(|\nabla\psi|^2 + \frac{R}{4}|\psi|^2 \right) + \int_M |D\psi|^2 = \int_{\partial M} \langle \nabla_\nu\psi + \nu \cdot D\psi, \psi \rangle \tag{13}$$

where ψ is a spinor and ν is the unit outer normal vector of the boundary.

The formula can also be proved by computing the divergence of a one form and applying Stokes' theorem. The specific one form α we use here is defined by:

$$\alpha(v) = \langle \nabla_v \psi + v \cdot D\psi, \psi \rangle$$

for $v \in TM$ and for a *fixed* spinor field ψ. The divergence of α is computed to be:

$$-\delta\alpha = |\nabla \psi|^2 - |D\psi|^2 + \frac{R}{4}|\psi|^2.$$

For a harmonic spinor satisfying $D\psi = 0$, the boundary integral on the right-hand side will be non-negative, provided the scalar curvature is non-negative. Moreover it can vanish if and only if ψ is globally parallel. To prove Theorem 13, we prove first the existence of a harmonic spinor which is asymptotically parallel in the sense that it approaches a parallel spinor (with respect to the background flat metric) sufficiently fast. The limiting value of the boundary integral is then shown to be the mass $(= E)$ when the boundary spheres go off to infinity. Rigidity follows from the fact that if the mass vanishes, we get a trivialization of the manifold by parallel spinors, since we get one for each asymptotic value.

To prove Theorem 14, one needs a connection $\widetilde{\nabla}$ that is flat for the standard hyperbolic space. There is a natural one, which I called a hyperbolic Cartan connection in [M1], which comes from imbedding hyperbolic space in Minkowski space and restricting the flat vector space parallelism. This can be done "virtually" for any Riemannian manifold on the stabilized tangent bundle $\widetilde{T}M = TM \oplus 1$, except that the connection would not be flat unless the manifold is hyperbolic.

More explicitly, the hyperbolic covariant derivative on the stabilized tangent bundle $\widetilde{T}M$ is defined by:

$$\widetilde{\nabla}_u v = \nabla_u v - \langle u, v \rangle e_0 \tag{14}$$

$$\widetilde{\nabla}_u e_0 = u \tag{15}$$

where $u, v \in \Gamma(TM)$, e_0 is the trivial section $e_0 : p \mapsto (0, 1) \in \widetilde{T}_p M = T_p M \oplus \mathbb{R}$, ∇ is the Levi-Civita connection, and $g = \langle, \rangle$ is the Riemannian metric on TM. ∇ and $\langle, \rangle$ can be extended to a Riemannian metric g on $\widetilde{T}M$ so that $g(e_0, e_0) = 1$, $e_0 \perp TM$ and $\nabla e_0 = 0$. This extension will be denoted by the same symbols g. On the other hand there is a natural Lorentzian metric $\tilde{g}$ on $\widetilde{T}M$, defined in the obvious way with $g(e_0, e_0) = -1$. This extension will be denoted by $\tilde{g}$. We have $\widetilde{\nabla}\tilde{g} = \nabla g = 0$ but $\widetilde{\nabla} g \neq 0$ and $\nabla \tilde{g} \neq 0$.

We point out here the obvious fact that in case M is hyperbolic space, the equations above are the Gauss equations for the standard imbedding of H^n into flat Minkowski space $\mathbb{R}^{1,n}$.

For spin manifolds, we also obtain induced connections on associated spinor bundles. For the spinor bundle $\widetilde{\mathbb{S}} = \mathbb{S}(\widetilde{T}M)$ the connection is given by:

$$\widetilde{\nabla}_u \psi = \nabla_u \psi - \frac{1}{2} e_0 u \psi \tag{16}$$

where $\psi \in \Gamma(\widetilde{\mathbb{S}})$, $u \in TM$ and the vectors $e_0, u \in \widetilde{T}(M)$ are acting on the spinor ψ via Clifford multiplication. The factor $\frac{1}{2}$ comes from the fact that $\mathrm{Spin}(n+1)$ is the double cover of $SO(n+1)$. The formulas for the connections are compatible in the sense that Clifford multiplication is parallel (i.e., $\widetilde{\nabla}$ satisfies the product rule). Moreover, the natural $\mathrm{Spin}(n,1)$-invariant Hermitian fiber metric on $\widetilde{\mathbb{S}}$ is $\widetilde{\nabla}$-parallel. The spinors $\widetilde{\mathbb{S}}$ associated to the extended group $\mathrm{Spin}(1,n)$ can be thought of as an appropriate number (depending on the parity of n) of copies of "ordinary" spinors $\mathbb{S}$ for $\mathrm{Spin}(n)$, where TM acts via Clifford multiplication as usual and where we introduce an extra element e_0 acting on $\widetilde{\mathbb{S}}$ with the following properties:

$$\langle e_0 \psi_1, \psi_2 \rangle = \langle \psi_1, e_0 \psi_2 \rangle, \quad e_0^2 = 1, \quad e_0 u + u e_0 = 0$$

for any $\psi_1, \psi_2 \in \Gamma(\widetilde{\mathbb{S}})$, $u \in TM$. Here $\langle,\rangle$ is the real part of the Hermitian inner product on $\widetilde{\mathbb{S}}$.

More specifically, we can take: $\widetilde{\mathbb{S}} = \mathbb{S}^+(TM) \oplus \mathbb{S}^-(TM)$ for *even* n and $\widetilde{\mathbb{S}} = \mathbb{S}(TM) \oplus \mathbb{S}(TM)$ for *odd* n. Under this identification, the Clifford multiplication of e_0 and of $u \in TM$ is given by:

$$\gamma(e_0) = \begin{bmatrix} 1 & 0 \\ 0 & -1 \end{bmatrix}, \quad \gamma(u) = \begin{bmatrix} 0 & u \\ u & 0 \end{bmatrix} \quad (\text{even } n),$$

$$\gamma(e_0) = \begin{bmatrix} 0 & -i \\ i & 0 \end{bmatrix}, \quad \gamma(u) = \begin{bmatrix} u & 0 \\ 0 & -u \end{bmatrix} \quad (\text{odd } n).$$

Again, we point out that for hyperbolic space H^n, $\widetilde{\nabla}$ is a flat connection and $\widetilde{\mathbb{S}}$ is trivialized by $\widetilde{\nabla}$-parallel spinors. These correspond, under the maps $(1 \pm e_0)$, to what are known as Killing spinors in the literature, i.e., if ψ is parallel $\widetilde{\nabla}\psi = 0$ then $\phi^\pm = (1 \pm e_0)\psi$ satisfies $\nabla_u \phi^\pm \mp \frac{1}{2} u \phi = 0$.

The Dirac operators corresponding to the two connections are defined by:

$$\text{(i)} \qquad D\psi = \sum_{k=1}^n e_k \nabla_k \psi \tag{17}$$

and

$$\text{(ii)} \qquad \widetilde{D}\psi = \sum_{k=1}^n e_k \widetilde{\nabla}_k \psi \tag{18}$$

where $\{e_1, \ldots, e_n\}$ is an orthonormal base for TM, and ∇_k means ∇_{e_k}.

We call $\widetilde{D}$ the *hyperbolic Dirac operator* in contrast to the "Euclidean" Dirac operator D. Using the super-commutativity of e_0 and D, $D e_0 + e_0 D = 0$, we obtain the following relation between the two Dirac operators:

$$\widetilde{D} = D + \frac{n}{2} e_0. \tag{19}$$

Using the identifications: $\widetilde{\mathbb{S}} = \mathbb{S}^+(TM) \oplus \mathbb{S}^-(TM)$ (even n) and $\widetilde{\mathbb{S}} = \mathbb{S}(TM) \oplus \mathbb{S}(TM)$ (odd n), we have:

$$\widetilde{D} = \begin{bmatrix} \frac{n}{2} & D \\ D & -\frac{n}{2} \end{bmatrix} \text{ (even } n\text{)}; \qquad \widetilde{D} = \begin{bmatrix} D & -\frac{n}{2}i \\ \frac{n}{2}i & -D \end{bmatrix} \text{ (odd } n\text{)}.$$

In particular, we note that $\widetilde{D}$ is a self-adjoint operator since Clifford multiplication by e_0 is symmetric. Since $D\,e_0 + e_0\,D = 0$, we obtain, upon squaring:

$$\widetilde{D}^2 = D^2 + \frac{n^2}{4}. \tag{20}$$

We make the important observation that $\widetilde{D}$ is a strictly positive elliptic operator and so there are no non-trivial L^2-harmonic spinors with respect to the hyperbolic Dirac operator.

Our next step is to derive the appropriate Lichnerowicz formula for $\widetilde{D}$. We first introduce the *rough Laplacians* defined by:

$$\nabla^*\nabla = -tr\,\nabla^2 = -\sum_{k=1}^{n} \nabla_k \nabla_k$$

$$\widetilde{\nabla}^*\widetilde{\nabla} = -tr\,\widetilde{\nabla}^2 = -\sum_{k=1}^{n} \widetilde{\nabla}_k \widetilde{\nabla}_k$$

where $\{e_k\}_{k=1,\ldots,n}$ is an orthonormal base for TM. $\nabla^*\nabla$ and $\widetilde{\nabla}^*\widetilde{\nabla}$ are both essentially self-adjoint, non-negative elliptic operators, and the relation between them is given by:

$$\widetilde{\nabla}^*\widetilde{\nabla} = \nabla^*\nabla + \frac{n}{4}. \tag{21}$$

This follows from the calculation:

$$\widetilde{\nabla}^*\widetilde{\nabla} = \left(-\nabla_k + \frac{1}{2}e_0\,e_k\right)\left(\nabla_k + \frac{1}{2}e_0\,e_k\right) = -\nabla_k\nabla_k - \frac{1}{4}(e_0\,e_k)^2 = \nabla^*\nabla + \frac{n}{4},$$

where we sum over an orthonormal base $\{e_k\}$.

The famous formula of Lichnerowicz for the ordinary Dirac operator is:

$$D^2 = \nabla^*\nabla + \frac{R}{4}$$

where R is the ordinary scalar curvature.

The corresponding formula for the hyperbolic Dirac operator $\widetilde{D}$ is therefore:

$$\widetilde{D}^2 = \widetilde{\nabla}^*\widetilde{\nabla} + \frac{\widetilde{R}}{4} \tag{22}$$

where $\widetilde{R} = R+n(n-1) = \sum_{j,k=1}^{n} \langle \widetilde{R}(e_j, e_k)e_k, e_j \rangle$ is the *hyperbolic scalar curvature*, since

$$\widetilde{\nabla}^* \widetilde{\nabla} + \frac{\widetilde{R}}{4} = \nabla^* \nabla + \frac{n}{4} + \frac{R}{4} + \frac{n^2}{4} - \frac{n}{4}$$

$$= D^2 + \frac{n^2}{4} = \widetilde{D}^2.$$

So we see that the analogue of Lichnerowicz' formula for the Dirac operator $\widetilde{D}$, defined by a hyperbolic Cartan connection is obtained simply by replacing the Riemannian connection and curvature terms by their hyperbolic analogues!

This is, in spirit, very similar to Witten's proof of the Lorentzian version (Theorem 12), where he also used the Levi-Civita connection of the surrounding space-time restricted to the space-like slice to define a modified Dirac operator. The curvature terms that appear in the Lichnerowicz formula for the square of Witten's Dirac operator involve more than just the scalar curvature. However, the dominant energy condition together with Einstein's equation is exactly what is needed to prove that the integrand is non-negative. The boundary integrals are identified, in the limit, with the total energy-momentum vector.

The proof of Theorem 14 now proceeds exactly as in the Euclidean case. First solve for a hyperbolically harmonic spinor with good asymptotics and then integrate by parts. We refer to [M1] for further details. Under the assumption that the manifold is strongly asymptotically hyperbolic, the analysis is in fact much more simpler than in the Euclidean space, since there is no need to introduce weighted Sobolev spaces to invert the elliptic operator $\widetilde{D}$.

The above proofs for positive mass theorems do not involve index arguments in contrast to results from the last section on compact manifolds. It would be interesting to have versions of these theorems for manifolds with (finite) boundary involving relative versions of the K-area and the index theorem. In fact, we can express the boundary integral purely in terms of a self-adjoint tangential Green operator acting on the boundary:

$$\nabla_\nu + \nu D = \nu \widehat{D} - \frac{H}{2} \tag{23}$$

where H is the mean curvature (with respect to the inner normal ν) of the boundary, and $\nu \widehat{D}$ is a tangential self adjoint boundary operator. We have then the following *fundamental formula* for a harmonic spinor:

$$\int_M \left(|\nabla \psi|^2 + \frac{R}{4} |\psi|^2 \right) = -\int_{\partial M} \langle \nu \widehat{D} \psi, \psi \rangle + \int_{\partial M} \frac{H}{2} |\psi|^2. \tag{24}$$

One would then impose appropriate boundary conditions to control the sign of the boundary integral.

3.3. Some Mathematical Aspects of the AdS/CFT Correspondence

There are many physical aspects about the AdS/CFT correspondence, more generally known as the holographic principle. I will only be able to describe a few special mathematical features, and my description is necessarily very limited.

First of all AdS stands for Anti-de Sitter space which is the Lorentzian analogue of hyperbolic space in Riemannian Geometry. CFT stands for conformal field theory and the correspondence is between supergravity (or string theory) of the bulk manifold (say, an asymptotically hyperbolic Einstein manifold) and conformal field theory on its boundary (which in the case of hyperbolic space is the conformal sphere). To be more specific, one studies a complete Riemannian Einstein manifold M^{n+1} with negative Ricci curvature, which has a conformal compactification in the following sense: M is the interior of a compact manifold with boundary $\overline{M}$ and the metric of M can be written near the boundary as:

$$ds^2 = \frac{1}{t^2}\left(dt^2 + g_{ij}(t,x)dx^i dx^j\right)$$

where t is a smooth function which is positive on M and has a zero of first order at $N = \partial \overline{M}$ (i.e., $dt \neq 0$ near N).

Although the induced metric on the boundary depends on the choice of the defining function t, the conformal class is independent of the choice and so conformal invariants on N are well defined. In particular, one is interested in expressing classical action functionals for (super-) gravity in terms of correlation functions of the boundary values at infinity. As a simple example, for scalar functions on hyperbolic space H^{n+1}, one can solve the Laplace equation for functions with any given prescribed boundary values. The resulting harmonic function is classically given by the Poisson kernel and so the Dirichlet functional (action) for harmonic functions can easily be written as a boundary integral on the conformal sphere S^n. (This is a two point correlation function for the boundary values). More generally one would study this correspondence for more elaborate functionals arising from supergravity and gauge theory on say, asymptotically hyperbolic spaces.

If the conformal class of the metric on the boundary contains a metric of positive scalar curvature (i.e., if the Yamabe invariant is positive), then Witten and Yau [WY] have shown that the conformal boundary N is connected, and also the nth homology group of $\overline{M}$ vanishes. This is a basic result, since the non-connectedness would have unnatural physical implications.

Somewhat more general results were then established by M. Cai and G. Galloway [CG]. The results can be formulated in a more geometrical fashion and the proofs are also based on more traditional comparison methods of Riemannian geometry.

Theorem 15. *Let M^{n+1} be a complete Riemannian manifold with compact boundary, suppose that the Ricci curvature of M satisfies $\mathrm{Ric}(g) \geq -ng$ everywhere, and suppose that the boundary has mean curvature $H > n$. Then M is compact.*

Theorem 16. *Let M^{n+1} be a complete Riemannian manifold admitting a conformal compactification $\overline{M}$, with boundary N^n, and suppose that the Ricci curvature of M satisfies $\mathrm{Ric}(g) \geq -n\,g$ everywhere and such that $\mathrm{Ric}(g) \to -n\,g$ sufficiently fast near the conformal boundary. Assume also that N has a component with a metric of non-negative scalar curvature. Then the following properties hold:*

(i) *N is connected.*

(ii) *If M is orientable then $H_n(\overline{M}, \mathbb{Z}) = 0$.*

(iii) *The map $i_* : \pi_1(N) \longrightarrow \pi_1(M)$ induced by the inclusion is onto.*

The Witten–Yau proof uses variational and comparison methods for minimal surfaces and related functionals for branes whereas Cai and Galloway use only variational and comparison methods for geodesics.

The proof of Theorem 15 is quite simple and follows from the fact that the condition $H > n$ on the boundary forces the geodesics starting perpendicular to the boundary to focus more strongly than the negative Ricci curvature of the bulk manifold, and hence these geodesics will have conjugate points within a finite distance. This is, of course, very reminiscent of the classical Bonnet–Myers argument. Theorem 16 is a bit harder to prove, but still is based on classical methods of Riemannian geometry using comparison arguments for Busemann functions. It would be nice to see a spinorial approach and maybe obtain stronger results using only lower bounds on the scalar curvature.

Finally, I would like to remark that there might be a connection of these type of problems with the entropy rigidity result of Besson–Courtois–Gallot [BCG]. In one of their proofs they used the fact that the imbedding of the manifold in the Hilbert space of all probability measures on the boundary at infinity given by the square root of the Poisson kernel calibrates the volume form. It is a homothety for the standard hyperbolic space. It would be interesting to extend this idea to other harmonic propagators, involving the Dirac operator, spinors, scalar curvature, K-area and the index theorem.

4. Epilogue

It has been almost half a century since my student days during the early 70s when I first read the seminal paper by Gromov and Lawson about scalar curvature on the torus; over three decades since I extended Witten's proof of the positive mass conjecture to the asymptotically hyperbolic case; and almost two decades since I lectured at the CRM in Castelló de la Plana in 2001 on those topics! Of course, there has been many new developments directly or indirectly related to the topics of these lecture notes since those days. A number of researchers have made improvements, generalized my theorems, and even disproved one of my conjectures! Since all of this was done independently (not in collaboration with me) and I was not involved, I will not be able to give a comprehensive survey here, but I will give a very brief selection instead.

4.1. Scalar curvature on the hemisphere

During the early 1990s, I optimistically announced the following conjecture about the scalar curvature rigidity of the hemisphere, after unsuccessfully attempting to prove it with spinorial methods using Cartan connections in the spirit of my proof in the hyperbolic case.

Conjecture. *Let M^n be a compact spin manifold with simply connected boundary ∂M and let g be a Riemannian metric on M with the following properties:*

(i) *∂M is totally geodesic in M;*
(ii) *the metric induced on ∂M by g has constant sectional curvature $K \equiv 1$;*
(iii) *the scalar curvature of g satisfies $R \geq n(n-1)$ everywhere on M.*

Then (M,g) is isometric to the round hemisphere with the standard metric.

This was proven to be false almost 20 years later, in 2011, by S. Brendle, F. Marques and A. Neves [BMN]. They have found a counter-example to my conjecture about the scalar curvature of the hemisphere.

Theorem 17. *Given any integer $n \geq 3$, there exists a smooth metric g on the hemisphere S_+^n with the following properties:*

(i) *The scalar curvature R_g of g satisfies $R_g > n(n-1)$.*
(ii) *At each point on ∂S_+^n, we have $g = \bar{g}$, where $\bar{g}$ is the standard metric on S_+^n.*
(iii) *The mean curvature of ∂S_+^n with respect to g is strictly positive (i.e., the mean curvature vector points inward).*

However my conjecture is true if one replaces the scalar curvature bound by a stronger bound on the Ricci curvature as was proved by F. Hang and X. Wang in 2009 [HW2] (it is also true if one restricts to the conformal class of the standard metric as was shown earlier in 2006 [HW1]).

Theorem 18. *Suppose that g is a smooth metric on the hemisphere S_+^n with the following properties:*

(i) *The Ricci curvature of g satisfies $\mathrm{Ric}_g \geq (n-1)$.*
(ii) *The induced metric on the boundary ∂S_+^n agrees with the standard metric $\bar{g}$*
(iii) *The second fundamental form of the boundary ∂S_+^n with respect to g has nonnegative principal curvatures.*

Then g is isometric to the standard metric on S_+^n.

My conjecture has led to a series of other papers (see [B] for a good review). These papers (unfortunately?) do not use Dirac operators and spinors. It would be intriguing to understand the correct boundary value problems for harmonic spinors that would lead to a better understanding of non-rigidity or rigidity of lower bounds on the scalar curvature.

4.2. Gromov's work on scalar curvature

Recently, Gromov has been able to generalize a number of results involving scalar curvature using minimal surfaces instead of the Dirac operator and spinors. This was done for low dimensions a while ago but very recently extended to all dimensions, based on new regularity results for minimal surface soap bubbles in higher dimensions established by Schoen and Yau [SY5]. In fact, Gromov has an extensive theory about how to look at scalar curvature form a much broader and more geometric point of view, beyond Dirac operators and spinors! In particular, Gromov laments the fact that one still doesn't have a good understanding of how the two methods relate in order to reconcile them and establish deeper connections. I would like to call the more "geometric" method using minimal surfaces and submanifolds, as was done by Schoen and Yau to prove the positive mass conjectures, the *covariant* approach and the more "physical" method using spinors and Dirac operators, as was done by Witten in his proof, the *contravariant* approach. My lectures are definitely in the second category! For more on Gromov's work we refer to the recent preprints [G3, G4, G5, G6]. The simplest and oldest example of these two approaches is the case of a positive lower bound for the Ricci curvature. On the one hand, one has the contravariant method used by Bochner to show vanishing of harmonic one forms, and hence of the first Betti number. On the other hand, one has the covariant method of analyzing the second variation formula for geodesics via Jacobi fields leading to the Bonnet–Myers theorem on the diameter, showing the compactness of the universal cover and hence finiteness of the fundamental group, which is a stronger result than the vanishing of the first de Rham cohomology group over the reals. The method of Schoen and Yau is based on a careful analysis of the second variation formula for minimal surfaces and the study of singularities that may arise. It is also interesting to note that many "vanishing theorems", such as that of Lichnerowicz or Bochner, prove the vanishing of all harmonic objects. This is, at least superficially, a lot stronger than just saying that some index of an elliptic operator is zero. The index is the asymmetry or disparity between harmonic objects of different parities (an anomaly, in physical lingo). A vanishing index only says that the two types of harmonic objects (zero modes) are of the same dimension. It doesn't say, like a vanishing theorem, that there are no zero modes at all. I find it intriguing that in many cases (BPS-states?) one type of zero modes is already automatically excluded, so the index is actually the dimension of harmonic objects a certain type. Another apparent advantage of the covariant method is the fact that it requires less smoothness, so perhaps it can deal with "rougher" metrics and even singularities. In fact, one of Gromov's goals is to have a more discrete and combinatorial understanding of scalar curvature for polyhedral objects.

4.3. Some speculations and musings

1. The traditional Atiyah–Singer Index Theorem relates the index of an elliptic operator to characteristic classes, which can be expressed locally as differential forms in terms of the curvature. This is best proved by evaluating the supertrace

of the heat kernel of the elliptic operator on the diagonal. Is there a version of the index theorem where the heat kernel is evaluated differently, for example outside of the diagonal, by letting one point go to infinity at a specific rate. One might not be interested in the classical topological index, but rather in the index density at the infinite (or finite) boundary. This is probably related to the Callias-type index and perhaps to the AdS/CFT correspondence (holographic principle).

2. Another speculation related to the Index Theorem is the fact that the local expansions of the heat kernel are always done with respect to the flat Euclidean metric as the background metric. Curvature is always measured with respect to Euclidean geometry, which by default is flat. My use of Cartan connections show that curvature can be measured with respect to other background metrics, especially that of symmetric spaces. It would be interesting to find an "index theorem" where the asymptotics of the heat kernel are calculated with respect to other background geometries, both locally and at infinity. The result might be more geometrical (in the spirit of AdS/CFT) than a purely topological statement about the index of an elliptic operator. One should also extend the Bochner–Weitzenböck formulas as I have done for the Lichnerowicz formula in the hyperbolic case.

3. The Euler characteristic can be localized around zeros of vector fields. Is there any way of localizing the $\hat{A}$-genus around codimension 2 submanifolds, relating it perhaps to the mean curvature? Is there a useful "transgression formula" for the $\hat{A}$-genus involving the scalar curvature, especially when $R > 0$? Some researchers have argued that positive scalar curvature should be reflected as positive Ricci curvature of the loop space. I would suggest that one should look at "secondary characteristics classes" and other cycles related to the Lie algebra of the diffeomorphism group (i.e., the Lie algebra of vector fields). This should also be useful to study Einstein metrics with negative scalar curvature. Speaking of Einstein metrics, one should perhaps also study the Rarita–Schwinger operator for $\frac{3}{2}$-spinors (which describe gravitinos in physics!), instead of just the Dirac operator.

4. Allow me to end with one last conjecture about the torus and nilmanifolds:

Conjecture. *The only Einstein metrics on compact nilmanifolds are the flat metrics on tori.*

References

[ACG] L. Andersson, M. Cai, and G.J. Galloway, "Rigidity and positivity of mass for asymptotically hyperbolic manifolds", *Ann. Henri Poincaré* **9** (2008), no. 1, 1–33.

[Ba]   R. Bartnik: "The mass of an asymptotically flat manifold", *Comm. Pure Appl. Math.* **39** (1986), 661–693.

[BGV] N. Berline, E.Getzler and M.Vergne: "Heat kernels and Dirac operators", *Grundlehren der math. Wiss.* **298** (1992), Springer-Verlag.

[BCG] G. Besson, G. Courtois and S. Gallot: "Entropies et rigidités des espaces localemant symétriques de courbure strictement négative", *Geom. Funct. Anal.* **5** (1995), 731–799.

[BH]   H. Boualem and M. Herzlich: "Rigidity at infinity for even-dimensional asymp-
        totically complex hyperbolic spaces", *Ann. Scuola Norm. Sup. Pisa* (Ser. V), 1,
        (2002) 461–469.

[Br]   H. Bray: "Proof of the Riemannian Penrose Conjecture using the Positive Mass
        Theorem", *J. Diff. Geom.* **59.2** (2001), 177–267.

[B]   S Brendle: "Rigidity phenomenon involving scalar curvature", *Surveys in Diff.
        Geom.* **XVII**(2012).

[BMN]   S Brendle, F.C. Marques and A. Neves: "Deformations of the hemisphere that
        increase scalar curvature", *Inventiones math.* **185.1**(2011) 175–197.

[BM]   S Brendle and F.C. Marques: "Scalar curvature rigidity of geodesic balls in S^n",
        J. Diff. Geom. **88** (2011) 379–394.

[CG]   M. Cai and G.Galloway: "Boundaries of zero scalar curvature in the AdS/CFT
        correspondence", *Advances in Theoretical and Mathematical Physics* 3(6) (2000).

[CH]   P.T. Chrusciel and M. Herzlich: "The mass of asymptotically hyperbolic mani-
        folds", *Pacific J. Math.* **212** (2003), 231–264.

[D]   H. Davaux, "An optimal inequality between scalar curvature and spectrum of the
        Laplacian" *Math. Annalen* **327.2**, (2003) 271–292.

[DM]   H. Davaux and M. Min-Oo: "Vafa–Witten bound on the complex projective
        space", arXiv:math.DG/0509034

[Go]   S. Goette: "Vafa–Witten estimates for compact symmetric spaces", Comm. Math.
        Physics **271.3**, (2007) 839–851

[GS1]   S. Goette and U. Semmelmann: "Spin^c structures and scalar curvature estimates",
        Annals of Global Analysis and Geometry **20.4**, (2001), 301–324.

[GS2]   S. Goette and U. Semmelmann: "Scalar curvature estimates for compact sym-
        metric spaces", *Differential Geom. Appl.* **16.1** (2002), 65–78.

[G1]   M. Gromov: "Positive curvature, macroscopic dimension, spectral gaps and higher
        signatures", Functional Analysis on the eve of the 21st century, vol. II, Progress
        in Math., vol. 132, Birkhäuser, Boston, (1996).

[G2]   M. Gromov: "Dirac and Plateau billiards in domains with corners", *Central Eu-
        ropean Journal of Mathematics*, Volume 12, Issue 8 (2014), pp. 1109–1156.

[G3]   M. Gromov: "Metric Inequalities with Scalar Curvature", *Geometric and Func-
        tional Analysis* **28.3**, (2018) 645–726

[G4]   M. Gromov: "Scalar curvature of manifolds with boundaries: natural questions
        and artificial constructions", arXiv:1811.04311.

[G5]   M. Gromov: "Mean curvature in the light of scalar curvature", arXiv:1812.09731.

[G6]   M. Gromov: "Four Lectures on Scalar Curvature", arXiv:1908.10612.

[GL1]   M. Gromov and H.B. Lawson: "Spin and scalar curvature in the presence of a
        fundamental group I", *Ann. Math.* **111** (1980), 209–230.

[GL2]   M. Gromov and H.B. Lawson: "The classification of simply connected manifolds
        of positive scalar curvature", *Ann. Math.* **111** (1980), 423–486.

[GL3]   M. Gromov and H.B. Lawson: "Positive scalar curvature and the Dirac operator
        on complete Riemannian manifolds", *Publ. Math. I.H.E.S.* **58** (1983), 295–408.

[GLS]   V. Guillemin, E. Lerman and S. Sternberg: "Symplectic fibrations and the mul-
        tiplicity diagrams", *Cambridge Univ. Press*, 1996.

[HW1] F. Hang and X. Wang, Rigidity and non-rigidity results on the sphere, Comm. Anal. Geom. 14, 91–106 (2006).

[HW2] F. Hang and X. Wang, Rigidity theorems for compact manifolds with boundary and positive Ricci curvature, J. Geom. Anal. 19, 628–642 (2009).

[He1] M. Herzlich: "Scalar curvature and rigidity of odd-dimensional complex hyperbolic spaces", *Math. Ann.* **312** (1998), 641–657.

[He2] M. Herzlich: "A Penrose-like inequality for the mass of Riemannian asymptotically flat manifolds", *Comm. Math. Phys.* **188** (1997), 121–133.

[Hi] N. Hitchin: "Harmonic spinors", *Advances in Math.* **14** (1974), 1– 55.

[HM] G.T. Horowitz and R.C. Myers: "The AdS/CFT correspondence and a new positive energy conjecture for general relativity", *Phys. Rev. D* **59** (1999), hep-th 9808079.

[HI] G. Huisken and T. Ilmanen: "The Riemannian Penrose inequality", *Internat. Math. Res. Notices* **20** (1997), 1045–1058.

[LM] H.B. Lawson, and M.-L. Michelsohn: "Spin Geometry", Princeton Math. Series 38, Princeton, 1989.

[Li1] M. Listing "Scalar curvature and asymptotically symmetric spaces", Commun. Math. Phys. 287, 395–429 (2009)

[Li2] M. Listing, Scalar curvature on compact symmetric spaces, arxiv:1007.1832

[Li3] M. Listing, "Scalar curvature and vector bundles", arxiv:1202.4325

[Ll1] M. Llarull: "Sharp estimates and the Dirac Operator", *Math. Ann.* **310** (1998), 55–71.

[Ll2] M. Llarull: "Scalar curvature estimates for $(n+4k)$-dimensional manifolds", *Differential Geom. Appl.* **6** (1996), no. 4, 321–326.

[Lo1] J. Lohkamp: "Metrics of negative Ricci curvature", *Ann. of Math.* **140** (1994), 655–683.

[Lo2] J. Lohkamp: "Curvature h-principles", *Ann. of Math.* **142.3** (1995), 457–498.

[Lo3] J. Lohkamp: "Scalar curvature and hammocks", it Math. Ann. **313** (1999), 385–407.

[Lo4] J. Lohkamp, The Higher Dimensional Positive Mass Theorem I, arXiv math.DG/0608795.

[Lo5] J. Lohkamp, Inductive Analysis on Singular Minimal Hypersurfaces, arXiv: 0808.2035.

[Lo6] J. Lohkamp, The Higher Dimensional Positive Mass Theorem II arXiv:1612.07505

[M1] M. Min-Oo: "Scalar curvature rigidity of asymptotically hyperbolic spin manifolds", *Math. Ann.* **285** (1989), 527–539.

[M2] M. Min-Oo: "Scalar curvature rigidity of certain symmetric spaces", *Geometry, topology, and dynamics* (Montreal), pp. 127–136, CRM Proc. Lecture Notes, 15, A.M.S.,1998.

[PT] T. Parker and C. Taubes: "On Witten's proof of the positive energy theorem", *Comm. Math. Phys.* **84** (1982), 223–238.

[Po1] L. Polterovich: "Gromov's K-area and symplectic rigidity", *Geom. Funct. Anal.* **6.4** (1996), 726–739.

[Po2] L. Polterovich: "Symplectic aspects of the first eigenvalue", *J. Reine Angew. Math.* **502** (1998), 1–17.

[RT]   O. Reula and K.P. Tod: "Positivity of the Bondi energy", *J. Math. Phys.* **25** (1984), 1004–1008.

[SY1] R. Schoen and S.T. Yau: "Existence of minimal surfaces and the topology of 3-dimensional manifolds with non-negative scalar curvature", *Ann. Math.* **110** (1979), 127–142.

[SY2] R. Schoen and S.T. Yau: "On the proof of the positive mass conjecture in general relativity", *Comm. Math. Phys.* **65** (1979), 45–76.

[SY3] R. Schoen and S.T. Yau: "On the proof of the positive mass conjecture in general relativity", *Manuscripta Math. Phys.* **65** (1979), 45–76.

[SY4] R. Schoen and S.T. Yau: "The energy and linear-momentum of space-times in general relativity", *Comm. Math. Phys.* **28** (1979), 159–183.

[SY5] R. Schoen and S.T. Yau: "Proof of the positive mass theorem II", *Comm. Math. Phys.* **79** (1981), 231–260.

[SY6] R. Schoen and S.T. Yau: "Positive Scalar Curvature and Minimal Hypersurface Singularities, preprint, 2017 arXiv:1704.05490 [math.DG]

[VW]   C. Vafa and E. Witten: "Eigenvalue inequalities for fermions in gauge theories", *Comm. Math. Phys.* **95** (1984), 257–276.

[Wa]   X. Wang "The mass of asymptotically hyperbolic manifolds", *J. Differential Geometry* **57** (2001) 273–299.

[W1]   E. Witten: "A new proof of the positive energy theorem", *Comm. Math. Phys.* **80** (1981), 381–402.

[W2]   E. Witten: "Anti-de Sitter space and holography", *Adv. Theor. Math. Phys.* **2** (1998), 253–290. hep-th/9802150

[WY]   E. Witten and S.T. Yau: "Connectedness of the boundary in the AdS/CFT correspondence", *Comm. Math. Phys.* **80** (1981), 381–402. hep-th/9910245.

[Wo]   E. Woolgar: "The positivity of energy for asymptotically anti-de Sitter spacetimes", *Classical and Quantum Gravity* **11.7** (1994), 1881–1900.

[Z]   X. Zhang: "Rigidity of strongly asymptotically hyperbolic spin manifolds", *Mathematical Research Letters* **7** (2000), 719–727.